Deep Seabed Resources

Deep Seabed Resources

Politics and Technology

Jack N. Barkenbus

THE FREE PRESS
A Division of Macmillan Publishing Co., Inc.
NEW YORK

Collier Macmillan Publishers
LONDON

The Free Press
A Division of Macmillan Publishing Co., Inc.
866 Third Avenue, New York, N.Y. 10022

Collier Macmillan Canada, Ltd.

Library of Congress Catalog Card Number: 78-73024

Printed in the United States of America

printing number

1 2 3 4 5 6 7 8 9 10

Library of Congress Cataloging in Publication Data

Barkenbus, Jack N.
Deep seabed resources.

Includes index.
1. Marine resources and state. 2. Ocean mining.
3. Manganese nodules. I. Title.
GC1017.B37 333.9'164 78-73024
ISBN 0-02-901830-7

Excerpts from some of the author's previously published works appear herein by permission of the publishers:

Excerpts, pp. 28–29 and 170, from "The Politics of Ocean Resource Exploitation" by Jack N. Barkenbus are reprinted from *International Studies Quarterly,* vol. 21, no. 4 (December 1977), pp. 675–700 by permission of the Publisher, Sage Publications, Inc.

Excerpt, pp. 155–160, from "Seabed Negotiations: The Failure of United States Policy" by Jack N. Barkenbus is reprinted with the permission of the San Diego Law Review Association from 14 *San Diego L. Rev.* 623 (1977), pp. 626–633.

Excerpt, pp. 173–174, from "The Future Seabed Regime" by Jack N. Barkenbus, originally published in the *Journal of International Affairs,* vol. 31, no. 1, Spring/Summer 1977, is reprinted with the permission of the Editors.

Excerpt, pp. 175–180, from "How to Make Peace on the Seabed" by Jack N. Barkenbus is reprinted with permission from *Foreign Policy* Magazine #25 (Winter 1976–77), Copyright 1976 by National Affairs, Inc, pp. 212–220.

To My Parents and My Wife, Belgin

Contents

Acknowledgments

This book has its origins in the two years (1973-75) the author spent as a member of the Center for Marine Affairs, Scripps Institution of Oceanography, LaJolla, California. Efforts by the staff of the Institute for Marine Resources at Scripps to educate our small group of social science landlubbers in the ways of the sea presented a unique challenge, and I am particularly grateful to Gerald Wick and my colleagues at the center, J. Ronald Stanfield and Paul Mandell.

The bulk of the book was written while receiving financial assistance from the Rockefeller Foundation under its Conflict in International Relations Program. I would like to thank John Stremlau for his efficient administration of the fellowship and his willingness to facilitate my research in any way he could. My research was conducted primarily in Washington, D.C., with my base at the School of Advanced International Studies (SAIS), The Johns Hopkins University (with the kind permission of Robert E. Osgood, dean of SAIS).

While at SAIS I was fortunate to be associated with its Ocean Policy Project, and I would like to thank the following for their helpful criticisms of earlier working drafts: Ann Hollick, James Orr, Kim Traavik, and Nathan Pelcovits. The cheerful administrative, editorial, and typing assistance of Barbara Bowersox was invaluable during my stay at SAIS.

I would also like to thank the editors and publishers of *Foreign Policy, San Diego Law Review, Journal of International Affairs,* and *International Studies Quarterly* for granting their permission to reprint sections from my earlier writings.

Finally, I would like to thank my wife for her patience and encouragement in the writing of this book.

Introduction

Nations are increasingly clashing over the division of the collective world product. Confrontations over access to, and the control of, resources, as well as the division of economic rent from the production and marketing of these resources, have become a persistent characteristic of the 1970s. There appears to be no imminent global reconciliation in sight which would prevent confrontations from also being characteristic of the 1980s. These confrontations arise fundamentally from the inability and/or unwillingness of industrial societies to satisfy their enormous requirements for resources from indigenous sources.

The questions of what constitutes a "fair" price for imported natural resources, and who will determine this value, are normally at the heart of the disputes. One generally finds industrial nations (the North) pitted against the nonindustrial nations (the South) in forums designed to answer these questions. Impetus for the confrontation was, of course, provided by the success of the Organization of Petroleum Exporting Countries (OPEC) in raising crude oil prices. Subsequently, developing nations banded together to forge the principles upon which a "new international economic order" (NIEO) would be based; these principles include, among others, the desire for restructuring producer-consumer trade agreements covering a wide range of commodities, and a universal recognition of "full and permanent" state sovereignty over natural resources. It is safe to say that little tangible progress has yet been made in these matters. However, there appears to be no diminution of the South's collective conviction that the fundamental tenets upon which economic exchange is now structured must change.

What is at stake, then, are contrasting, collectively held visions of the future. Disputes that might appear amenable to pragmatic conciliation are now being consciously placed within the framework of these strongly held preferred visions, and in the process vastly complicate the arrival at mutually satisfactory solutions.

This book is about one such dispute. It illustrates that even when mining enterprises go to the earth's last frontiers–i.e., the seabeds in the deepest sections of the oceans–they cannot avoid the pervasive global politics involving resources today. In a very real sense, the political dispute over the deep seabed resources–manganese nodules–is a microcosm of the larger global struggle between North and South. It is replete with familiar conflicts over the role and autonomy of multinational corporations, the exploitation and supply of raw materials, the use of such standards as economic efficiency and economic equity, and the apportionment of voting rights among member nations of international organizations. Because it is so representative of the international struggle for economic control taking place today, it possesses significance far in excess of its immediate substantive importance. This study's analysis goes beyond the narrow economic issues at stake and examines the dispute within the framework of competing ideologies.

Commercial mining of manganese nodules has yet to commence, and the dispute over who will mine and how these resources will be mined in the future has been the major stumbling block to the successful formulation of a comprehensive law-of-the-sea treaty. Nations have been attempting to forge consensus on this issue for the past decade—first, within the U.N. Seabed Committee, and later, within the U.N. Conference on the Law of the Sea (UNCLOS). The lack of UNCLOS success in producing an acceptable legal framework for the mining of manganese nodules has severely tested the patience of those mining enterprises developing the requisite technology and the patience of government representatives from several nations, including the United States. The year 1979 may, in fact, witness the international community's last chance to resolve this issue through global negotiations. Unilateral mining of manganese nodules could very well begin within a year, which could in turn lead to open conflict on the high seas.

The primary features distinguishing the manganese nodule dispute from other resource-related issues at UNCLOS are, first, that the vast majority of nodules are found on the seabed beyond even extended national jurisdictions; second, because they belong to no nation, manganese nodules have generally been recognized to be the "common heritage of mankind"; and third, the international community several years ago committed itself to the establishment of an international organization to oversee the mining effort. The deadlock at UNCLOS has been over the creation of this international organization, to be termed the International Seabed Authority (ISA). Perceived North and South interests diverge over defining ISA's powers, functions, and tasks and establishing the proper internal decision-making structure of the organization itself. The major purpose of this book is to explore and explain why negotiations over the ISA have progressed so unsatisfactorily and to offer a compromise formula in which all interests could be satisfied over time. The book also intends to demonstrate that any attempts by the United States to bypass the UNCLOS forum and proceed with unilateral mining will produce results inimical to U.S. goals.

Since manganese nodules—normally no more than 4 inches in diameter—are found on top of the seabed at ocean depths of 12,000 to 20,000 feet, there has been little reason until recently for people other than scientists to be aware of their existence. This situation has changed, of course, because of the advance of technology. Although it was known some time ago that manganese nodules contained minerals essential to an industrial society, it was not until the 1960s that major organizations determined that these minerals could be extracted at a profit. This belief is still be be confirmed since commercial manganese nodule mining has yet to begin; but the many mining firms (based in several industrial nations) active in technological development and testing today remain optimistic.

Were politics not a part of manganese nodule mining, the subject would still be of considerable fascination. Man's search for materials essential to sustaining industrialization has now taken him to the furthest depths of the oceans and involve a unique technological enterprise. In an earlier age, such a significant endeavor would have gained universal approbation. Things are no longer so simple.

This book brings an international perspective to the dispute over manganese nodules, but it focuses on U.S. interests and policy-making. Mining firms within the United States and the government itself have played major, if not determining roles in the dispute. Moreover, pressure to bypass UNCLOS negotiations and support unilateral mining emanates primarily from segments of the U.S. government (as well as from domestic mining organizations). The government, therefore, has been at the center of the controversy and is likely to remain a key actor in the future.

The book contains three major parts and a conclusion. The first part (Chapters One through Five) describes the various major dimensions of the dispute, providing the reader with the background necessary to approach analysis of international negotiations. Chapter One gives technical detail, and Chapter Two deals with the legal ambiguities currently facing deep seabed mining. Chapter Three assesses who will benefit from the mining of manganese nodules. Chapter Four identifies the changing nature of mining in general, and places manganese nodule mining specifically within these changing trends; and, finally, Chapter Five reveals the conflicting positions within the U.S. government itself concerning this issue.

Part Two (Chapters Six through Eight) elaborates upon UNCLOS negotiations and examines the major issues in dispute. Chapter Six discusses varying national positions concerning the degree of supranationality with which the ISA is to be endowed. In other words, the functions and powers of the prospective ISA are examined. Chapter Seven outlines the dispute over which nations will be appointed to positions of power within the ISA, and Chapter Eight probes the controversy over the specific mining conditions under which mining enterprises will exploit manganese nodules. Possible means of structuring the relationship between existing mining entities and the prospective ISA are discussed.

Part Three (Chapters Nine and Ten) is devoted to explaining the manganese nodule dispute in terms of the broad and contrasting economic ideologies of North and South. It concludes that nations have actually been talking past one another in the UNCLOS forum, rather than attempting to find a common solution through negotiations. Chapter Nine reveals how U.S. negotiators have approached the UNCLOS forum and why this approach has been unsuccessful.

Chapter Ten examines the relationship of the nodule issue to the South's call for NIEO. The intransigence of the South at UNCLOS is shown to be a product of the collective negotiating posture increasingly being demonstrated in the international arena.

Finally, Chapter Eleven offers for consideration a possible basis for accommodation which could, if accepted, bridge the positions of opposing forces. It would require compromises from both North and South, but would prevent the anarchy at sea to which the present course of action is heading.

The major thesis of this book is that unilateral mining of manganese nodules, in direct opposition to the wishes of most states, will, on the one hand, gain the United States very little and, on the other hand, place much at risk. For a decade now, U.S. administrations have been committed to resolving the seabed dispute within an international framework. Unfortunately, they have not been as equally committed to seeking a truly international solution. As a result of the intransigence of all parties, disillusionment with UNCLOS has become commonplace, and its annual, ten-week sessions may soon come to an end without producing the desired treaty. It is impossible to predict what would happen should negotiations be terminated without a resolution of the issue. Those who naively expect the problem to disappear would, no doubt, be disappointed. The most tragic aspect of UNCLOS' demise, however, would be the lost opportunity to demonstrate that nations can, in response to the increasing interdependence brought about by technology, fashion a stable and equitable world order. If it cannot be done within the confines of UNCLOS, the prospects for its demonstration elsewhere are not good.

PART 1

The Policy Context

Chapter 1

Resources, Technology, and Economics

The dispute over the mining of manganese nodules is perhaps unique in that the resources in question are characterized by abundance (not scarcity) and exploitation has yet to commence. The dispute, therefore, is over how mining is to proceed in the future and not over any current inequities. This chapter essentially deals with three facets of the prospective mining enterprise: the resource base; the technologies involved, as well as the organizations developing them; and the major economic parameters.

Oceanographers have known of manganese nodules for approximately a century, but the possibility of their commercial exploitation has been realized only relatively recently. This chapter highlights the role of a single pioneer in bringing manganese nodules to the attention of established mining companies. Whether, and at what pace, mining will proceed depends fundamentally on technological feasibility and economic factors. This chapter details the technical uncertainties surrounding the various mining system components, such as the mining device, the lifting device, surface vessels, ore transport, and processing. The organizations involved in technical research and development are also revealed.

Finally, this chapter deals with the economic uncertainties that make manganese nodule mining a high-risk endeavor, all political considerations aside. It is shown that although the mining companies involved present an optimistic picture of commercial feasibility, it is still too early to reach such a conclusion.

Resources

Much of the concern of some observers over possible depletion of resources stems from a limited and faulty use of this concept. Too often the concept is focused on known and commonly used materials which serve man today in some manner. A broader and more appropriate definition of resources would stress not only substances but man's knowledge and capabilities. Resources, therefore, are the product of man's ingenuity and technology. This view, often termed the "functional concept of resources,"[1] is far less pessimistic regarding depletion, since man's ingenuity can produce resources where none existed before.

At present manganese nodules cannot be deemed a resource, as man has yet to use them in the service of his wants and needs. It is only the recent development of deep seabed mining technology that has led to their imminent retrieval and the vision of their widespread utilization. We are on the brink, therefore, of creating, not discovering, a vast storehouse of resources through technological innovation.

Up to this time, manganese nodules have been solely scientific curiosities, largely unknown throughout the world because of their concentration on top of the most inaccessible section of the globe's surface, i.e., the seabeds located in the deepest parts of our oceans. It has only been a hundred years since even scientists have been aware of their existence. With the advent of modern oceanography in the nineteenth century, serious exploration of the great oceans began. One of the major milestones in early oceanographic exploration was the British expedition of the HMS *Challenger* in the 1870s. As the *Challenger* went from one ocean to another, scientists dredged the ocean bottom for samples. Much to their surprise, the scientists discovered that there was on the seabed an abundance of either brown or black potato-shaped particles, varying in size and shape but averaging approximately 4 inches in diameter.

After the properties of these very porous nodules were assayed (revealing the presence of thirty elements surrounding nuclei, such as fish bones and shark teeth), scientists aboard the *Challenger* found the question of their origin and the process of their formation a difficult puzzle. There was considerable speculation regarding their origins but no long-term research effort was embarked upon, and the *Challenger* collection essentially collected dust in British museums for years. The early 1950s brought a revival of interest, however, until today it is a rather lively center of scientific debate with numerous competing theories being intensively examined.[2] One researcher has identified the following areas of scientific interest now being pursued:[3]

1. delineation of the source of metals (volcanism, ambient sea water, diffusion from sediments) found in manganese nodules
2. circulation of hydrothermal waters in igneous rocks of ridge zones
3. quantitative estimates of age, and solution of the paradox relating "older" nodules to young host sediments
4. primary and secondary features of nodules and their bearing on the history of nodule formation and environment
5. detailed distribution of highly metalliferous nodules and also microdistribution of metalliferous components
6. mineralogy of nodules and physiochemical study of mineral stability
7. diffusion modeling of manganese migration in sediments
8. physical oceanographic and geological setting of nodules (including sediment transport)

Research in the United States up to this time has been primarily supported by the National Science Foundation's International Decade of Ocean Exploration Program. It is not manganese nodules as scientific phenomena, however, that is the topic of this book or that has instigated a major international dispute. International conflict does not attend disputes over scientific validity but does, all too often, characterize those over resources.

The essential reason for global interest in manganese nodules resides in their abundance. They have been found in all oceans (even in some lakes), and estimates of their aggregate weight run into trillions of tons. Since less than 5 percent of the total ocean floor has been surveyed in any detail, it is impossible to place a reliable figure on the absolute magnitude of the abundance. One study claims that there are approximately 1.5 trillion tons of nodules in the Pacific Ocean alone.[4] It has been estimated that 80 to 90 percent of the nodules lie on the seabed beyond even the furthest claims to national jurisdiction, i.e., 200 miles from any land. There is substance in the claim of oceanographers, therefore, that the nodules potentially comprise the largest mineral deposit on this planet. This is unquestionably a very unique deposit, as it extends horizontally rather than vertically and is covered by water rather than soil—and best of all, it is a deposit yet to be tapped by man.

In an age increasingly concerned with material shortages, manganese nodules represent an anomaly. Rather than depletion, we are faced with abundance—which has been reassuring to at least one observer. Robert Heilbroner, who in his book, *An Inquiry Into the Human Prospect,* painted a bleak picture of man approaching finite resources, later modified his assessment: "I would take a more hopeful view of the mineral resources problem. This is because I did not know when I wrote the book of the possibility of mining the seabeds for mineral nodules which are estimated to exist on the order of hundreds of billions of tons."[5]

As will be pointed out throughout this volume, despite their abundance, one should not, as Heilbroner apparently does, view manganese nodules as setting us free from material constraints. Manganese nodules have no application in the area where depletion is now most serious, namely, energy. Moreover, the difficulties that will face at least first- and second-generation miners in raising them from the seabed should not be underestimated. In other words, the creation of manganese nodules as a resource should be reassuring, but there is no point served in overestimating or exaggerating their importance in the course of human events.

Manganese nodules contain approximately thirty elements, as seen in Table 1, but only four are of commercial interest at this time: copper, cobalt, nickel, and manganese.

Copper is, of course, a versatile mineral for which there is currently a very high demand. Because of its electrical conductivity and corrosion resistance, it is most extensively used in the electrical and allied industries. The other three desired minerals, cobalt, nickel, and manganese, are all primarily steel alloys. Nickel, for example, is often used in stainless steel applications. Nickel-alloy

TABLE 1. Pacific Manganese Nodules Weight Percentages (Dry Weight Basis): Statistics on 54 Samples

ELEMENT	AVERAGE	MAXIMUM	MINIMUM
Manganese	24.2	50.1	8.2
Iron	14.0	26.6	2.4
Silicon	9.4	20.1	1.3
Aluminum	2.9	6.9	.8
Sodium	2.6	4.7	1.5
Calcium	1.9	4.4	.8
Magnesium	1.7	2.4	1.0
Nickel	.99	2.0	.16
Potassium	.8	3.1	.3
Titanium	.67	1.7	.11
Copper	.53	1.6	.028
Cobalt	.35	2.3	.014
Barium	.18	.64	.08
Lead	.09	.36	.02
Strontium	.081	.16	.024
Zirconium	.063	.12	.009
Vanadium	.054	.11	.021
Molybdenum	.052	.15	.01
Zinc	.047	.08	.04
Boron	.029	.06	.007
Yttrium	.016	.045	.033
Lanthanum	.016	.024	.009
Ytterbrium	.0031	.0066	.0013
Chromium	.001	.007	.001
Gallium	.001	.003	.0002
Scandium	.001	.003	.001
Silver	.0003	.0006	......

SOURCE: Congressional Research Service, *Ocean Manganese Nodules* (Prepared for the Senate Committee on Interior and Insular Affairs, Washington, D.C., Feb. 1976), p. 5.

steels are used in the production of pollution-control equipment, electroplating, and jet engines and turbines. Cobalt adds to the heat resistance of steel. For this reason, it is often used in the production of high-speed tools and jet aircraft. Because of its important magnetic and chemical properties, cobalt can be used to produce electromagnets and heat-resistant turbine blades.

Managnese is a highly versatile mineral for which there is much demand in industrial societies. Its foremost use, again, is as a steel alloy. Thirteen to 20 pounds of manganese alloyed with a ton of steel can make the final steel product more resistant to shock and abrasion. Manganese is also used as a scavenger in steel, reducing levels of sulfur, oxygen, and trace impurities. The most common form of manganese is called ferromanganese, which is both cheap and abundant. The manganese from the nodule, on the other hand, is high-purity manganese, which is likely to be more expensive than ferromanganese to extract and, therefore, probably will not replace ferromanganese in its steel applications—hence the reluctance on the part of many nodule miners to consider marketing manganese. One mining firm, Deepsea Ventures, does plan to market it, however, but not as a steel alloy. Instead, manganese from the nodule would compete in the "premium price market," which consists of essentially low-volume uses and high prices. A possible future use of manganese is as an additive to gasoline in place of lead, which would reduce auto emissions. Because of its scavenger properties, manganese could be used in many other different ways to improve the environment.

It is quite clear, therefore, that manganese nodules can provide minerals essential to the functioning of an industrial society. Of course, it is conceivable that new uses of the elements found in the nodules will be discovered in order to exploit their abundance, but it is impossible to predict such a development. It is also possible that the sediments of the deep ocean seabed may, in the future, be prized more than the manganese nodules that rest atop it. These sediments include red clay, which could be a possible source of aluminum, and various oozes such as calcium carbonate, which could substitute for limestone in the production of cement. However, this book is devoted to current uses and applications of manganese nodules primarily because political debate is focused upon such application.

The Initial Entrepreneur

The possible transformation of manganese nodules from a scientific curiosity to a valuable resource was fully grasped first by one man: John L. Mero. A scientist with entrepreneurial interests and inclinations, Mero first proclaimed the manganese nodule a potential resource in an article published in 1952.[6] Subsequently, he participated in a cooperative project between the Department of Mineral Technology at Berkeley and the Institute for Marine Resources (located at Scripps Institution of Oceanography) to examine the extent of seabed minerals and possible means for their recovery. It was in his 1959 report, "The Mining and Processing of Deep-Sea Manganese Nodules" (Institute for Marine Resources), emanating from this project that he first set forth what he believed to be the technical and economic parameters of deep seabed mining.

Since that time, Mero has been an indefatigable promoter of manganese nodule mining, leaving the academic community to convince others of the viability of the enterprise. For several years, he found few people and organizations interested. His economic projections, which appear extremely optimistic today, have always been seriously questioned. Mero, himself, maintains that manganese nodule mining has always been a profitable investment but that the larger mining companies, not wishing to see alternative and competing sources of ore developed, preferred to ignore him and seabed deposits. A more likely explanation is simply that mining companies were extremely reluctant to take the word of a lone pioneer about such a novel and speculative enterprise. Some mining companies did initiate programs in the early 1960s—such as Kennecott Copper and Deepsea Ventures—but these initial efforts were simply to survey the resource base, not to initiate technological development. In addition, talk in the early 1960s of a "wet NASA", where government would commit the same level of funding to explore "inner space" as it had outer space, never materialized. Mero's enthusiasm for getting on with the job, therefore, was not reciprocated. Neither the government nor private industry was willing to make major initial commitments on the basis of Mero's technical and economic analysis.

Over the years, it has become apparent that Mero is not adverse to mixing scientific fact with hyperbole. In many of his presentations he claims that manganese nodules are forming in the Pacific Ocean at a rate faster than they can be mined and consumed,[7] giving the misleading impression that these are indeed renewable resources. In fact, manganese nodules grow extremely slowly, perhaps at the rate of 2 to 4 millimeters per *million* years. Mero may be technically correct that tons of new nodule material are forming, in the aggregate, because of the sheer abundance of existing nodules. What he does not explain, and which will be examined shortly, is that it is extremely unlikely that most nodules will ever be of commercial interest. Consequently, the growth on these nodules is of no interest to us whatsoever. The growth occurring on the nodules of commercial interest is relevant, however, but it is insignificant in the time frame pertinent to man.

Mero has also been cavalier in dismissing the possibility of any negative environmental impact that might result from deep seabed mining. He did not address this issue in his early works, and indeed it did not really surface until the late 1960s and early 1970s in association with the heightened American environmental consciousness. Since 1970, possible environmental damage occasioned by the mining and processing of manganese nodules has been a lively area of debate and scientific research. Our ignorance of the organisms that exist on or above the deep seabed, their ecological niche, and general oceanic processes is simply so great that dispute over environmental impact probably will continue for some time. It may, in fact, delay actual commercial exploitation if and when legal and political disputes among nations are resolved. In recognition of this fact, the U.S. government has in the past few years devoted considerable resources to generating a base of data relevant to the environmental issue which can aid in the analysis of potential problems.[8] In any event, Mero seems oblivious to the fears of environmentalists and actually claims that manganese nodule mining will result in a positive environmental impact—through the closing down of environmentally damaging land-based mines (since he claims that land-based mines will not be able to compete with manganese nodule mining and consequently will be shut down within twenty years).[9]

The major fundamental issue, however, remains that of technical and economic feasibility. As stated previously, Mero claims that it always has been possible to mine manganese nodules at an economic profit and that no new technical innovations are necessary. In his 1965 book he states, "Calculations and laboratory experiments indicate that there should be no major problems in adopting existing industrial equipment and processing to the mining and processing of manganese nodules."[10] Mero gives the impression that only a few minor land-based mining adaptations in technology are required to make the enterprise a success. It is difficult for a nonengineer to assess the validity of his claim. Does mining at such depths represent a discontinuity from current practices? Does it represent a technological breakthrough of major proportions? Speaking with engineers and examining their writings one gets divergent impressions. Those who agree with Mero marshall the following arguments.

In terms of hard technology, all one needs is a scale markup of existing technology. The same principles currently being applied in dredges, for example, can be applied to deep seabed dredges. In addition, offshore oil development has produced much technology that can be transferred directly to the mining of hard minerals. In order to stabilize oil exploration ships, special instrumentation and devices were developed for the oil industry. These developments also can be used in deep seabed mining. Likewise, the oil companies needed long pipes with which to drill into the sediments in search of oil and gas. Manganese nodule mining will require long pipes as well. In terms of hardware, therefore, much of what is being used today on a smaller scale can be adapted by the manganese nodule industry, and there is much to be gained from the experience of the oil industry.

Furthermore, the movement from land to ocean mining has its advantages in terms of extraction. Water, not earth, is the overburden covering the desired resource, and consequently no blasting or extensive drilling is required. Moreover, there are no drafts to drive or shafts to sink as in conventional mining. Since the deposit is horizontal rather than vertical, one need only gather nodules, not extract them. Finally, instead of expensive drilling for land surveying, television cameras can survey a complete manganese nodule deposit, eliminating the uncertainty as to its extent.

In summary, one can argue that manganese nodule mining represents no new level of technical achievement but simply an extension of technologies we already have. On the other hand, there are those, in contrast to Mero, who insist that such mining will represent a technological feat of major proportions. They bring a good deal of skepticism with them, in fact, as to whether seabed mining can be carried out profitably in the near future.

Sorensen and Mead pointed out in a 1968 article[11] that dredges being used at this time were designed for work in only 90 feet of water and that the operation of dredges at depths of 2 to 4 miles represented a totally unique undertaking. Even more central, the pipe extending from the ship 2 to 4 miles down to the seabed would be subject to considerable stress because of the varying pressures of the ocean depths. It was recognized that the construction of a durable and reliable pipeline would require engineering of the highest order. As Deepsea Ventures president, Jack Flipse, confided in 1976, some colleagues thought he was "weak-minded" ten years previously for even contemplating such a difficult technical feat.[12] Some, no doubt, still feel that way.

The impression left by the arguments cited previously was that the ocean represents a more desirable mining environment or locale than land. Such an impression can certainly be challenged. Anyone who has either lived or worked on the oceans for an extended period of time will attest to the unpredictable and hostile environment that must be endured. Because of our lack of sustained working experience in such an environment, such problems as ocean corrosion can represent major obstacles. There is a widely held public misconception that the seabed is flat. In fact, it contains peaks which measured from the bottom are taller than Mt. Everest. Generally, though, the seabed is undulating. As a scientist for one of the mining companies has explained, where flat areas of the seabed do exist, they are generally devoid of significant manganese nodule deposits.[13] Designing dredges to operate on these slopes, therefore, is a difficult task, and it is estimated that perhaps 20 to 25 percent of a mine site may be inaccessible because of topographical obstacles. Thus the impression that nodules are sitting on a horizontal plane simply waiting to be plucked needs to be countered.

Thus far, only the ocean extraction process has been discussed, but as we will see later, processing is every bit as important to the successful operation of mining as actual exploitation. Prior to serious research and development there was no processing method known to be the most desirable means of extracting the important elements from the nodule. Considerable subsequent work in this area has proceeded, but not yet beyond the pilot project stage. Full-scale processing, therefore, represents another technical hurdle to surmount.

It should be evident now that one can adopt both optimistic and pessimistic views about the possibility of venturing into this commercial mining enterprise. Very little insight can be gained from the mining projects currently involved in research and development since much of their work is proprietary. No company wants to detail the problems they are having and the proposed solutions for fear that their competitors will find useful information in these revelations.

The technical components of the mining system will be described shortly in detail; but before leaving this subject another point should be raised concerning why so little is being revealed about the magnitude of this effort, to wit, the politics of the situation. Mining companies have fervently insisted that it is not technical problems that are holding up commercial mining but the legal-political climate. As Marne A. Dubs, director of the ocean resources department of Kennecott Copper, has stated in congressional testimony, "There is no longer any doubt about the technical and economical feasibility of ocean mining. The technology is ready; the investment climate is not."[14]

Clearly Dubs is presenting a far too sanguine picture of technological development. Were the technology so clearly in hand, there would be little need for companies to embark upon the ambitious precommercial testing now being planned. The purpose in making such definitive pronouncements is to put pressure on the appropriate political representatives to resolve the legal-political obstacles. It is not the case, however, that favorable resolution of the political dispute will necessarily bring forth a healthy and thriving ocean mining industry. What is clear is that unless the political dispute is resolved, there will be very little testing of mining systems to ascertain whether a viable industry can exist. The concern of mining companies is understandable. The impression they deliberately give, however, of a full-blown industry being restrained by legal-political factors is quite misleading and contributes to the "crisis" mentality often evident when policy-makers approach this issue.

Although Mero did not deal with environmental issues in his early writings, he was quite aware of the legal ambiguity surrounding exploitation of the nodules. He did not realize that this problem would be of such large proportions, but he certainly cannot be criticized for that—few could have predicted the events that would occur fifteen to twenty years later. In a 1960 article he even proposed U.N. involvement in formulating an international ocean mining law.[15] This proposal was to be forcefully and eloquently expanded upon some seven years later.

This section has touched briefly upon the projected technologies that will be used. It is necessary, however, in the following section, to outline in some depth what constitutes the ocean mining enterprise.

Ocean Mining Technology

Mining is often said to consist of five separate stages of operation: prospecting, exploring, research and development, prototype development, and commercial exploitation. American firms involved in seabed mining have yet to progress beyond the first three stages. One reason for this lack, of course, is that major investments are involved in the latter two stages, and companies are waiting for a more favorable investment climate before devoting large sums of money. Probable figures for the respective stages in the ocean mining enterprise might be $20 to $30 million for the first three stages combined; $50 to $100 million for the fourth prototype development; and approximately $600 million for the final stage, commercial exploitation. It is evident, therefore, that U.S. companies have invested relatively little up to this point in terms of the possible total requirement.

American universities and their primary sponsor of research within the federal government, the National Science Foundation, played a major role in the initial prospecting stage. Scientists, particularly those from the Lamont-Doherty Geological Observatory, contributed substantially to identifying the distribution of manganese nodules throughout the ocean seabeds. Mining companies have found these data useful in their work but have had to do more intensive investigation of the deposits—the exploration stage—on their own. Flipse of Deepsea Ventures classified his company's projected activities beyond the exploration stage as follows:[16]

Research (through 1974):
- locate mine site
- test components
- test mining system
- pilot process
- identify products
- estimate economics

Development and Evaluation (1975–1977):
- file claim, develop mine
- dredge tonnage with on-site mining machinery
- produce tonnage metals from demonstration plant
- market test products
- reverify cost/revenue projections

Construction (1978–1980):
- complete mining plant
- construct mining and transport system
- construct process plant
- negotiate sales agreements

Commercial (1981–):
- operate system
- explore for richer deposits
- expand mining and processing systems
- make profit

It is clear that major expenditures will accompany the latter two stages. It is less clear whether the dates for the latter stages listed above will be met.

Commercial Exploitation

Commercial exploitation is made up of five separate activities or components: the mining device, lifting method, surface vessel, ore transport, and processing. It is necessary to elaborate upon each component to provide a composite picture of the entire enterprise.

Developing the mining device has itself been an activity strongly pushed in the past, and according to Deepsea Venture officials, only within the last two years has this been developed to their satisfaction.[17] The reason a good deal of emphasis has been placed upon this element of the enterprise is that the efficiency with which nodules are collected can mean the difference between profit and loss.

The dredge head is the most common mining instrument being developed to collect nodules. The major problem involved is designing a dredge that can collect and handle different sizes. Because the lifting unit is only a certain diameter, the dredge must be equipped with a screen to sort out nodules that are too large.

Passive dredge heads (in contrast to self-propelled ones) that must be dragged along the bottom by the mining pipe at approximately 2 to 3 miles per hour are now being tested. A passive dredge loses the flexibility inherent in a self-propelled model but is far less expensive. A television camera can be placed on the dredge to give advance notice of the terrain and the field of nodules.

The mining devices in the continuous line bucket (CLB) system are, of course, the buckets attached to the line or rope. These buckets are made of steel and are dragged across the seabed to scoop up the nodules. The bucket is probably even less efficient than the dredge, but it does have an advantage in that it can recover nodules of varying sizes. Commercial buckets are likely to have a 5-ton capacity.

The second major component of a deep seabed mining system will be a lifting device. That of the hydraulic system consists of string pipe, which when carried on board is in segments but when operational can extend 4 miles deep into the oceans. There are two possible means of bringing the slurry of water and nodules to the surface. The method favored by the International Nickel Company (INCO) and Kennecott Copper uses a conventional centrifugal pump, wherein pumped water provides the necessary upward flow. It is a well-tested method of raising objects and is used extensively in coal mining today. The difficulty with this method is that the pump must be kept in the water either near the seabed or at intermediate depths—thereby subject to the stress of ocean weather, corrosion, and other problems of the ocean environment. The other hydraulic method, based on injecting compressed air into the pipe at various depths, is the system currently preferred by Deepsea Ventures and some foreign firms. The positive feature of this system is that the pump remains on board; the negative feature is that it is perhaps less efficient than the other method and that the three-phase flow may be more complex than one based on pumped water.[18]

The lifting device of the CLB system—a system pioneered by the French and Japanese—is the continuous loop of high-strength polypropylene rope, with buckets attached at regular intervals. With only traction drive on the mining ship to move the cable and lift the buckets, this system is far more simple than the hydraulic one. It is by no means free of problems, however, as snags in the cable may develop, and the cable may even snap on occasion under the tension being applied. The CLB system has already been used to some extent in offshore commercial tin production.

Two other lifting systems have been designed but are not now being tested for use in manganese nodule mining. The Summa Corporation reportedly planned to position a submersible barge between the mining device and the long pipe, which would transport the slurry back to the mining ship. The barge would function as an undersea launching platform from which the mining device or collector would be employed. The barge, upon receiving the nodules, would crush and clean them before sending them on their way up the pipe. With the barge directing and propelling the mining device, greater efficiency presumably would be achieved.

The other designed lifting system is perhaps the most imaginative one yet proposed, in which the mining device and lifting mechanism are combined in lunar-type robot vehicles.[19] The vehicles, or capsules, would be programmed to descend to the ocean floor, gather the nodules mechanically up to their lifting capacity, and then return to the mining ship. The originators of this design envisioned a total of ten vehicles, each making approximately eight trips per day, each one bringing back a total of 30 tons of nodules per day. The large mining concerns developing the commercial technology have not yet taken this scheme seriously.

The mining vessels themselves will of necessity be large, particularly those using the hydraulic lifting method, because they must be capable of storing the extensive equipment (primarily the long segments of pipe) and also must contain a storage area for recovered nodules. As mentioned before, precise positioning of the mining vessel will be essential, and it will be required to work in heavy seas 365 days of the year. A conveyor system able to transfer nodules from the mining vessel to ore carriers will also be a part of the overall mining system. It is likely that each company involved in mining will have to keep a fleet of ore carriers operating between sea- and land-based operations. The size of such carriers is likely to range anywhere from 35,000 to 60,000 tons deadweight.

Methods for processing manganese nodules were under experimentation in the late 1960s. Because of their special character, normal means of processing land-based minerals (by either physical separation or smelting techniques) have not proved satisfactory. There have been many methods devised to beneficiate and refine the ore by chemical separation processes. Details regarding each method are not publicly available since the information is deemed proprietary. It is known that Kennecott Copper currently favors an ammoniacal leaching process, which separates the nickel, copper, and cobalt from the nodule but not the manganese. At the present time, Kennecott does not plan to market manganese. One way designed to extract all four of the minerals is through a hydrochlorination leaching process. Nodules are crushed, and through the use of hydrogen chloride at high temperatures the minerals are extracted. Other possible routes to processing include the use of sulfuric acid and sulfur dioxide. It should be recognized that processing techniques in practice will be extremely sensitive to the specific concentration of minerals found in the recovered nodules—which means that nodules from one geographic area may not be substituted interchangeably with others while being processed. In other words, the requirements of processing dictate a site-specific approach to nodule mining. All processing plants at the moment are being planned at land-based sites, except for Lockheed Missiles and Space, which is looking seriously at ocean-going processing.

Who Is Involved?

One study has claimed that, globally, there are over one hundred companies active in developing manganese nodule mining technology.[20] Although this may well be the case, it gives the mistaken impression that the field is already very crowded and competitive. In fact, the vast majority of companies have simply invested a nominal amount in order to cautiously investigate greater involvement in the future. As such, the deep seabed mining field is dotted with consortia, a well-known method of sharing both risks and costs. Only a very few companies have forcefully and boldly made the level of investment and dedication necessary to be the acknowledged leaders in the field. Table 2 describes the five major consortia that have been formed to date.

The companies involved in technology development are based in eight nations: Australia, Belgium, Canada, the Federal Republic of Germany, France, Great Britain, Japan, and the United States. There are no third-world companies now involved in this field, and as we will see later, this fact has significant political bearing. Although the above consortia all contain a mix of national identities, we may see a more national orientation develop as commercialization becomes more imminent.

There are two American-based firms which have been involved in technological development since the early 1960s and which are acknowledged to have a technological lead; these are Deepsea Ventures and Kennecott Copper. Deepsea Ventures, unlike Kennecott Copper, was formed specifically to mine manganese nodules. It began corporate life in 1962 as the Newport News Shipbuilding and Dry Dock Company and is based in Glouchester, Virginia. During the early and mid-1960s, it conducted several prospecting expeditions, and in 1968 became Deepsea Ventures with the financial support of Tenneco Oil Company (the relationship with Tenneco was terminated in 1975). Deepsea Ventures has probably been the most active firm in exploration and technological development. In the late 1960s it conducted numerous tests of the technology off the east coast of the United States and on the Blake Plateau at depths of from 2,400 to 3,000 feet. Atlantic Ocean nodules appear to be of far less commercial interest than Pacific Ocean ones, however, and the company's theater of operations has shifted to the Pacific. Deepsea Ventures was also the first company to announce a breakthrough in processing methods; in 1971 it confirmed that a hydrochloric acid leaching method would be used for mineral extraction. Subsequently, a 1-ton-per-day pilot processing plant was constructed in Virginia to test and develop this method. Unlike most other firms, Deepsea Ventures plans to extract, process, and market the manganese contained in the nodules. Besides forming a consortia to continue research and development, it recently bought an ore carrier and began converting it to an ocean mining vessel.

Kennecott Copper has not "gone public" with its nodule program to the same extent as Deepsea Ventures. It is known, however, that the company began prospecting and technological development as early as 1962. Although details are not known, company spokesmen claim that their mining equipment has been tested at depths of up to 15,000 feet. Kennecott has an experimental processing lab at San Diego and a half-ton-per-day processing pilot plant at Lexington, Massachusetts.

It should be stated that only one company, Deepsea Ventures, has announced

TABLE 2. Ocean Mining Industrial Involvement

CONSORTIA/COMPANIES	INVESTMENT SHARE (%)	NATIONALITY	DESCRIPTION
Ocean Mining Association:			
Deepsea Ventures	service contractor	U.S.	Total $20 million R&D program from 1976 through 1979
U.S. Steel	33 1/3	U.S.	
Union Minière	33 1/3	Belgium	
Sun Oil	33 1/3	U.S.	
Kennecott Consortium:			
Kennecott Copper	50	U.S.	Total $50 million R&D program from 1976 through 1981
Rio-Tinto Zinc	10	U.K.	
Consolidated Gold Fields	10	U.K.	
Noranda Mines	10	Canada	
Mitsubishi	10	Japan	
British Petroleum	10	U.K.	
Ocean Management Group:			
International Nickel (INCO)	25	Canada	Began preliminary mining tests in 1978
Arbeitsgemeinschaft Meerestechnische gewinnbare Rohstoffe (AMR)	25	Federal Republic of Germany	
Deep Ocean Mining (DOMCO)	25	Japan	
SEDCO	25	U.S.	
Ocean Minerals Co:			
Lockheed Missiles and Space	25	U.S.	Contracted the *Glomar Explorer* in 1978 for ocean mining tests
Standard Oil of Indiana	25	U.S.	
Royal Dutch Shell	25	U.K./Netherlands	
Bos Kolis Westminster	25	Netherlands	

TABLE 2. Ocean Mining Industrial Involvement

CONSORTIA/COMPANIES	INVESTMENT SHARE (%)	NATIONALITY	DESCRIPTION
French Association for Nodule Exploration:			
Centre National pour l'Exploitation des Océans (CNEXO)	unknown	France	Primarily involved in site exploration
Commissariat a l'Énergie Atomique (CEA)	....	France	
Bureau de Recherche Géologiques et Minières (BRGM)	....	France	
Société Metallurgique Nouvelle / Société Le Nickel (SMN/SLN)	....	France	
Chantiers de France Dunkerque	....	France	
Continuous Line Bucket Group:			
Approximately 20 companies from six countries: Australia, Canada, France, Federal Republic of Germany, Japan, and the U.S.			Formed to develop the CLB technology; not for mining commercialization

SOURCE: Data from Congressional Research Service, *Deep Seabed Minerals: Resources, Diplomacy, and Strategic Interests* (March 1, 1978), p. 15; *idem, Ocean Manganese Nodules* (June 1975), pp. 35–36; Jessica Mott, "Fact Sheet on Seabed Mining," *Neptune,* No. 10 (May 1977), p. 4.

where it intends to build a full-scale processing plant (in Belgium). According to Kaufman of Deepsea Ventures there are two key criteria for the selection of processing sites: the availability of energy and environmental protection standards.[21] Because of the abundance of activities competing for energy in this country and the strict environmental standards, it may very well be that all processing plants will be established outside the United States.

Another group that had its origins in the pre-1970 period is the CLB group. This conglomerate has periodically conducted tests in the Pacific, dating back to 1968. Although the results of these tests have not been commented on in detail, some observers have claimed that they have not gone well at all and that the basic technology is still very much undeveloped.[22]

It appeared for several years that the Summa Corporation (owned by Howard Hughes) would be the first mining entry in commercial production. The real and bizarre mission of Summa's "mining vessel," the *Glomar Explorer,* was revealed in 1974 by the U.S. press. Summa's professed mining interests, it seems, were simply a cover to obscure its goal of recovering a sunken Soviet submarine. The budding industry was completely unaware of Summa's covert mission before the press revelations, although there was considerable talk about the enterprise before 1974. One competitor claimed early in 1974 that Summa was building such a sophisticated mining system that it would be too expensive to compete (claiming that Hughes was buying "a Cadillac of a system").[23] Speculation on what Hughes was up to often turned to the possibility of Summa selling its technology rather than becoming involved in the actual mining operation.

Although secrecy has been a hallmark of the industry for obvious commercial reasons, the extent to which Summa went to ensure it did not go unnoticed. Requests for information by the U.S. government's National Oceanic and Atmospheric Administration (NOAA) went unanswered. *Ocean Mining News,* an ocean technology newsletter, dubbed the Summa enterprise "the Great Hughes Ocean Mining Mystery," not because they suspected a covert mission but because of the large-scale effort and the unprecedented attempts to impose secrecy. Security guards and barbed wire were in evidence wherever technological development was being undertaken. The Hughes empire served as a perfect cover for the real mission, as the unprecedented secrecy was attributed to Hughes' personal eccentricities. A 1973 *Business Week* article on the adventure stated, "People in the mining industry have little doubt as to what Hughes is up to: He is putting $250 million into the *Glomar Explorer* and its submersible barge so he can scoop up the manganese nodules."[24]

There has been no attempt by Summa Corporation to go into manganese nodule mining since the revelations. It has been decided, for one thing, that the mining vessels belong to the U.S. government and not to the Summa Corporation. Efforts by U.S. government officials to get Deepsea Ventures, Kennecott Copper, or Lockheed Missiles and Space interested in leasing the *Glomar Explorer* and ancillary technology were unsuccessful until 1978, when Lockheed consented to use the ship over a period of two years.

In summary, it is largely incorrect to claim that nations as such possess the potential capabilities to exploit manganese nodules. At the moment, nongovernmental organizations are taking the lead in this enterprise, and as we have seen in the formation of consortia, national affiliations are essentially insignificant. The extent and form of government support to nongovernmental mining organizations vary considerably from country to country. In some countries, such as Japan, there is a strong government-corporation relationship whereby considerable government investment is provided. In the United States, on the other hand, the government has not been intimately involved in the development of technology. It is not publicly known if any mining entities from socialist nations are involved in technological development. The Soviet Union has conducted several prospecting missions, but whether a decision has been made to actually pursue mining in the future is unknown. The Soviets are generally self-sufficient in the minerals found in the nodule, and therefore, would presumably have less incentive to support mining than would other countries.

Mining Economics

The composition of elements within manganese nodules varies considerably from one seabed site to another. Although some regions of the Pacific seabed are known to contain nodules of certain characteristics, the variability even within regions is significant. The varying concentration is crucially important because it is widely believed that the mining of manganese nodules with average concentrations of the desired minerals will not be profitable. This belief is based on the expected competition such mining is likely to face from land-based mining. It is not enough, therefore, for mining companies simply to locate an abundant field of nodules. Unless these nodules are "high grade" (i.e., having higher than average concentrations of the desired minerals), they will be of little interest. It is, of course, impossible to place a precise figure on what nodules will or will not be economical to mine. There are many factors both endogenous and exogenous to seabed mining that can affect these considerations. Nevertheless, Deepsea Ventures scientist Siapno gave what he considered to be a minimally commercial grade of ore as 20 percent manganese, 1 percent nickel, .8 percent copper, and .2 percent cobalt.[25] These are minimum grades, and the search will be for much richer nodules. Since companies have expressed most interest thus far in extracting copper and nickel, the search is essentially for either copper-rich or nickel-rich nodules. An attractive mining site, for example, has been described as one with nodules containing a combined nickel-copper content of at least 2.5 percent.[26]

It is difficult at this time to predict with certainty where the "high-grade" nodules are located. Exploration has been most intensive in the Pacific Ocean, and the highest grade is thought to exist in the large area southeast of Hawaii falling between 0° and 20° north latitude in the range of 120°–140° west longitude—particularly in a narrow band perhaps 200 kilometers across and 1,500 kilometers long, running east-west around 9° north latitude. Even within this area there is apparently no discernible pattern that can predict the concentration of high-grade nodules.

Hence, although there are trillions of tons of nodules in the oceans of the globe, the gathering or exploiting of these potential resources is a good deal more complicated than one would initially expect. The question, then, is not simply can one raise enough nodules to make a profit, but can one raise enough *high-grade* nodules to make a profit. This requirement focuses attention upon an important mining principle, namely, the necessity of obtaining long-term assured access to specific mining sites. Some observers, unfamiliar with mining economics and procedures, have suggested that mining companies simply "fish" for manganese nodules, and thereby not obtain concessions to a specific site. Because of the variability of manganese nodule content, however, such a proposal is unsatisfactory. Mining companies, despite the acknowledged abundance of nodules, cannot afford to mine without some guarantees of exclusive access. Such guarantees not only enable exclusive access to high-grade nodules, but also enhance the possibility of mining nodules that are consistent compositionally—the latter feature being essential for efficient processing of nodules. "Fishing" for nodules would admittedly lessen the political problems inherent in this pioneering venture, but the suggestion ignores the very real and relevant considerations that make it unacceptable to mining companies.

What constitutes seabed reserves and resources, then, has yet to be determined. The term "reserves" generally refers to minerals that can now be extracted profitably with existing technology and under present economic and legal conditions. "Resources" is a more encompassing category of minerals that may in the future become reserves, given changes in technical, legal, or economic circumstances. First-generation miners will be interested in reserves rather than resources. The fact that there are trillions of tons of nodule resources means little to those attempting to make a profit now. One industry representative has estimated that seabed reserves of copper and nickel might be equal in quantity to those on land.[27] If this is indeed the case, manganese nodules comprise a body of ore which, although vast, is something less than the accounts given in the popular press.

The simple, fundamental fact is that there will be no manganese nodule industry unless mining can meet the imperatives of the proverbial "bottom line." It is tempting to view the legal-political imbroglio that has ensnarled the budding industry for the past ten years as the single factor responsible for holding back mining. However, even in the most favorable political-legal climate, manganese nodule mining would still represent a very speculative and high-risk enterprise. It would indeed be ironic if the international community were to provide a favorable legal consensus upon which to structure seabed mining only to find it still too risky to attract more than token investment. Developments are taking place in land-based mining (e.g., the possible mining of laterite deposits for nickel) that may be a more attractive investment than seabed mining. There is no assurance, therefore, that we will see large-scale mining of the seabed in the foreseeable future. It has been noted that many of the companies now investing in early research and development are doing so, not because they view the seabed as an attractive investment, but because they find it a necessary "defensive" strategy. In other words, in order to be prepared for unfortunate contingencies that may arise in a company's land-based operations, it is always wise to have an alternative source of mineral supply. Companies may not view the seabed as their preferable theater of operations but instead as a prudent alternative to their preferred locations. If this is the case, seabed mining is likely to progress in a far more incremental fashion than now envisioned.

Instead of dealing in generalities as we have thus far, it is important to present a few of the economic factors or characteristics that will distinguish seabed mining as an enterprise. The purpose of this forthcoming section is not to provide a careful and detailed examination of the economics of the industry. There already have been many studies, initiated outside the mining industry itself, that attempt to come to grips with detailed economic projections.[28] The utility of these studies, I submit, is not in the microanalysis purporting to determine the extent of the industry's return after investment. There are far too many unknowns at this time to give a precise, credible range. Moreover, there is far too much proprietary information, which limits the ability of outsiders to make feasible estimates. Instead, what can be usefully identified are the characteristics that set this pioneering venture apart from others and color the political positions adopted by the companies involved.

The first major economic characteristic of the manganese nodule industry having significant implications is the fact that there are extremely high front-end capital costs. That is, before a single nodule is commercially marketed, firms must raise large sums of capital. (In this sense, ocean mining is really no different from mining in general.) To obtain a reasonable return on investment, therefore, successful operation over an extended period of time is required.

How high initial capital costs will be has been a subject of much speculation. The estimates made in the 1960s now appear significantly below mid-1970 estimates, even when one takes inflation into account. Table 3 gives an estimate of the range of costs–capital and operating–that companies are likely to incur.

The capital costs (initial capital and working capital) alone for one mining venture run from a possible low of $425 million to a high of $600 million, which is a far cry from the Mero estimate in 1960 of $100 million.[29] Most ensuing studies, particularly from the 1968–1974 period, showed capital costs ranging from $200 million to $250 million.[30] It seems, therefore, that there has been a significant re-evaluation of capital costs since 1975 and that estimates continue to suggest spiraling costs. This rise has been confirmed by several industry executives in their testimony before congressional committees. Operating costs are expected to be significantly lower than capital costs, in a range (seen in Table 3) running from $120 million to $165 million.

TABLE 3. Ocean Mining Cost Estimates (1975)

	LOW	MEDIUM	HIGH
Exploration and R&D	$ 75[1]	$125	$150
Capital Costs	$385	$468	$550
Total Investment	$460	$593	$700
Working Capital	$ 40	$ 45	$ 50
Annual Operating Costs	$120	$143	$165

SOURCE: Rebecca L. Wright, "Ocean Mining: An Economic Evaluation," Ocean Mining Administration, Department of the Interior, May 1976, p. 11.
[1] Millions of 1975 dollars.

High capital costs have been at the center of the controversy surrounding the formation of this new industry. The inability of companies to produce such large funding from their own accounts requires that they obtain it from banks. Banks, however, have been unwilling to commit such funds for this purpose, given the legal uncertainties that abound. The risks and the sums are deemed too large for a prudent banking investment. Thomas C. Houseman, a vice-president of Chase Manhattan Bank, in testimony before a Senate subcommittee made his position very clear: "In view of the demonstrated desire of the international community to establish control over such activity, the present absence of political sponsorship and security of tenure constitute an unacceptable business risk to a financial institution."[31]

The inability of mining companies to obtain the large funds necessary to initiate the commercial phase of deep seabed mining effectively stops them from moving ahead. It is unclear whether most companies, if not all, have yet reached a point at which they want to commit major funds even if they could. The longer the delay in establishing a definitive legal code, however, the greater the chance of reaching such a point. It is possible for the U.S. government to provide mining companies with investment guarantees, thereby eliminating the risk involved to the companies and banks. Houseman stated that were the companies to receive investment guarantees in legislation, financial support from the banks would be forthcoming.[32] (Further elaboration of this possibility can be found in Chapter Five.)

The second major economic characteristic is one generally not recognized; namely, the glamour of the enterprise may be in the operations taking place in 15 to 20 thousand feet of water, but the majority of expenses will be in the processing plant and other land-based operations. Much has been made of the advantages and disadvantages associated with competing mining systems, e.g., the CLB versus the hydraulic systems. The former will require much less capital investment than the latter, but it is not expected to raise as many nodules. Which system will ultimately prove more efficient remains to be determined. The point is, however, that the economics of the total enterprise will still be determined as much, if not more, by land-based operations as by any sea-based system. Hence, although CLB advocates talk about reduced capital requirements, they have yet to find a means of processing that is significantly less expensive than any others.

Operating expenses will be contingent, in part, upon the quantity of nodules raised. Such firms as Deepsea Ventures, which will process and refine all four of the major minerals of interest, are aiming for the recovery of at least 1 million short dry tons per year. Other firms, which are planning to refine cobalt, copper, and nickel but not manganese, are aiming for the recovery of 3 million short dry tons per year. Companies claim such quantities may be required to turn a profit, but as a whole, operating cost estimates vary widely. The range of estimates goes from $18 per ton to a high of $75 per ton.[33] Dubs of Kennecott Copper has provided a most optimistic picture of operating costs; he estimates that mining recovery will constitute $5 to $10 per ton, transportation, $4 to $7 per ton, and processing, $10 to $15 per ton—a total of $19 to $32 per ton of nodules recovered. Dubs also estimates that the sale of the three minerals—copper, cobalt, and nickel—will bring $65 to $75 per ton recovered, which would leave a return on investment (before taxes) of anywhere from $33 to $56 per ton recovered. He further claims that companies might obtain $10 to $15 per ton after accounting for taxes and risks incurred.[34] Were such figures believed within the mining industry, Kennecott would no doubt have more competition than it now faces. The Dubs figures are even more optimistic than those of Mero some fifteen years ago when he predicted total operating costs of from $35 to $50 per ton recovered.[35] The analysts from Deepsea Ventures give a far less optimistic estimate, projecting mining and transportation costs at $20.5 to $25 per ton and processing at $34.5 to $50 per ton.[36] Therefore, operating costs, according to these estimates, will range from $55 to $75 per ton.

Since the range of operating costs is so large at this time, very little can be said definitively. It should be repeated, however, that land-based processing will be a major cost component.

The third major economic characteristic is the large impact of outside forces on mining profitability. These forces, which for the most part are outside the capabilities of the miners themselves to affect, can determine the extent of manganese nodule mining in the future. Three key exogenous factors are (1) the particular share of economic rent to accrue to international and/or national authorities, (2) the future prices at which nodule minerals will be sold, and (3) the competitiveness of nodule mining with land-based mining.

It has long been recognized that mining is a risky enterprise. Mining companies try to reduce risks in all their operations, and for this reason it is more accurate to characterize large mining firms as "risk minimizers" rather than "profit maximizers." In other words, caution and stability are viewed as desirable attributes rather than the reckless pursuit of a bonanza. Deep seabed mining is no exception, and several companies have worked methodically to reduce the considerable technological risks that are involved. They have worked through national representatives to the United Nations Conference on the Law of the Sea (UNCLOS) to reduce the political risks extant. Placing their fate in the hands of politicians, however, keeps them from having direct control over events, and there has been considerable fear on the part of industry officials that government representatives will not produce the desired investment climate. The inconclusive nature of UNCLOS proceedings has not reassured them. The general investment climate, therefore, and the extent of economic rent that will accrue to government authorities is outside the ability of the companies to resolve. Most mining concerns have committed themselves, in principle, to some form of revenue-sharing or profit-sharing with an international authority. As will be described later, there are several schemes now being considered as the search for an appropriate basis continues. What would possibly be objectionable for many companies would be the imposition of national taxation as well. Double taxation could present an obstacle to many companies who are now considering the pros and cons of getting involved. Wright has pointed out that government taxes could make the difference as to whether investment will or will not move into the area.[37]

One cannot predict what the market price of nodule minerals will be when they first reach the market. As is well known, mineral markets are very volatile. For example, the price for copper skyrocketed during 1973 and 1974, but since that time it has returned to pre-1973 levels. Nodule mining will also be subject to such fluctuations. There are now underway, of course, intensive negotiations on mineral prices in general in an effort to create a system of exchange not subject to the boom-bust cycle now prevalent. Should a more stable and predictable price formula be agreed upon between buyers and sellers, the uncertainties involved in ocean mining would be somewhat reduced. Chapter Three will examine the potential impact of nodule mining itself on mineral prices.

Finally, one cannot forget that nodule mining will have to compete with land-based mining. Unlike the market for seabed hydrocarbons, the market for the minerals found in the nodule is not unlimited. Nor does there appear to be imminent exhaustion of land-based sources of these minerals. Investment in seabed mining will indeed be limited if it proves to be a high-cost source, and as mentioned previously, laterite mining may prove to be a more promising investment for nickel exploitation. Another factor that could affect mineral prices is substitution. One significant development, for example, is the potential replacement of copper in long-range telecommunications with glass—which could lower the price of copper. One cannot simply assume, therefore, that the advent of nodule mining will bring a flood of investment.

Conclusion

This chapter has dealt with the economics of nodule mining in only a cursory fashion, primarily because estimates at this time are so speculative that detailed analysis is hardly fruitful and is subject to limitless qualification. Until mining is actually underway, there will be little basis for confident prediction. The author feels that seabed mining will progress very slowly at least throughout the end of this century. The technological uncertainties are considerable, and time will be needed to produce a fully reliable and resilient system. Moreover, as we will see, the political uncertainties, which are beyond the direct influence of mining companies, are considerable and not likely to disappear in the near future. The transformation of nodules from a scientific curiosity to a valuable resource will probably, therefore, be slower than many now perceive.

Notes

1. William N. Peach and James A. Constantin, *Zimmerman's World Resources and Industries,* 3rd ed. (New York: Harper & Row, 1972), p. 534.
2. Edward D. Goldberg, "Marine Geochemistry, 1. Chemical Scavengers of the Sea," *The Journal of Geology,* 62, 3 (May 1954), 249–65.
3. F. T. Manheim, "Manganese Nodules: Basic Science and Applications," Paper presented at the National Science Foundation's Seabed Assessment Workshop, July 16, 1975, p. 3.
4. Congressional Research Service, *Ocean Manganese Nodules,* Prepared for the Senate Committee on Interior and Insular Affairs, Washington, D.C., Feb. 1976, p. xiii.
5. Robert L. Heilbroner, "Second Thoughts on the Human Prospect," *Challenge* (May/June 1975), p. 24.
6. John L. Mero, "Manganese Nodules," *North Dakota Engineer,* 27 (1952), 28–32.
7. John L. Mero, "Potential Economic Value of Ocean-Floor Manganese Nodule Deposits," in David R. Horn, *Ferromanganese Deposits on the Ocean Floor,* Arden House Conference, January 20–22, 1972 (Washington, D.C.: National Science Foundation), p. 195.
8. The U.S. Department of Commerce, National Oceanic and Atmospheric Administration, has had for a number of years a continuing assessment of the environmental impact from deep seabed mining under the heading, the Deep Ocean Mining Environmental Study (DOMES).
9. As cited in "Manganese Nodules: Prospects for Deep Sea Mining," *Science,* 183 (Feb. 15, 1974), 646.
10. John L. Mero, *The Mineral Resources of the Sea* (Amsterdam: Elsevier, 1965), p. 277.
11. Philip E. Sorensen and Walter J. Mead, "A Cost-Benefit Analysis of Ocean Mineral Resource Development: The Case of Manganese Nodules," *American Journal of Agricultural Economics,* 50, 5 (1968), 1614.
12. Statement of John E. Flipse cited in *The Science, Engineering, Economics and Politics of Ocean Hard Mineral Development,* 4th Annual Sea Grant Lecture and Symposium, MIT Sea Grant Program, Oct. 16, 1975, p. 12.

13. W. D. Siapno, "Exploration Technology and Ocean Mining Parameters," in *Current Developments in Deep Seabed Mining,* Hearings before the Senate Minerals, Materials, and Fuels Subcommittee, 94th Congress, 1st Session, Nov. 7, 1975, p. 232.
14. *Ibid.*, p. 4.
15. John L. Mero, "Minerals on the Ocean Floor," *Scientific American* (Dec. 1960), p. 71.
16. John E. Flipse, "Ocean Mining–Its Promises and Its Problems," *Ocean Industry,* 10, 8 (Aug. 1975), 134.
17. Statement of Richard Greenwald before the Committee on Economic Potential of the Ocean, Marine Technology Society, Washington, D.C., Jan. 19, 1976.
18. *Science,* "Manganese Nodules," p. 644.
19. Bernard G. Stechler and John T. Nicholas, "Recovery of Deep Ocean Nodules: A New Approach," in *Ferromanganese Deposits on the Ocean Floor,* pp. 141–48.
20. Congressional Research Service, *Ocean Manganese Nodules,* p. 37.
21. Raymond Kaufman, "Land-Based Requirements for Deep-Ocean Manganese Nodule Mining," in *Manganese Nodule Deposits in the Pacific,* Symposium/Workshop Proceedings, State of Hawaii, Oct. 16–17, 1972, p. 113.
22. *Mineral Resources of the Deep Seabed,* Part 2, Hearings before the Senate Minerals, Materials, and Fuels Subcommittee, 93rd Congress, 2nd Session, Mar. 3–11, 1974, p. 1081.
23. "Manganese Nodules," *Science,* p. 645.
24. *Business Week* (June 16, 1973), p. 47.
25. Siapno, "Exploration Technology," p. 237.
26. United Nations, *Economic Significance, In Terms of Seabed Mineral Resources, of the Various Limits Proposed for National Jurisdiction: Report of the Secretary-General* (A/AC. 183/87), June 4, 1973, p. 19.
27. *Mineral Resources,* p. 1063.
28. See, for example, Nina W. Cornell, "Manganese Nodule Mining and Economic Rent," *Natural Resource Journal* (Oct. 1974), pp. 519–32; Sorensen and Mead, "A Cost-Benefit Analysis," pp. 1611–20; Rebecca L. Wright, "Ocean Mining: An Economic Evaluation," Ocean Mining Administration, Department of the Interior, May 1976; David B. Brooks, "Deep Sea Manganese Nodules: From Scientific Phenomenon to World Resource," *The Future of the Sea's Resources,* Proceedings of the 2nd Annual Conference of the Law of the Sea Institute, June 26–29, 1967, pp. 32–41.

30. Ta M. Li and C. Richard Tinsley, "Meeting the Challenge of Material Demands from the Oceans" *Mining Engineering* (April 28, 1975), p. 29.
31. In *Current Developments in Deep Seabed Mining,* p. 13.
32. *Ibid.*, p. 19.

33. Li and Tinsley, "Meeting the Challenge," p. 29.
34. *Ocean Science News* (May 17, 1974), p. 4.
35. Mero, "Minerals on the Ocean Floor," p. 67.
36. Arnold J. Rothstein and Raymond Kaufman, "The Approaching Maturity of Deep Ocean Mining–the Pace Quickens," *Mining Engineering* (April 1974), p. 53.
37. Wright, "Ocean Mining," p. 16.

Chapter 2

The Legal Background

Before seabed mining beyond national jurisdictions can proceed, a new and widely recognized legal regime for this exploitation must be established. The effort to do so was initiated in 1967 by the United Nations. From 1968 through 1973, this attempt took place within the United Nations Seabed Committee and since that time under UNCLOS auspices. This chapter is devoted to explaining why many nations and, indeed, the miners themselves feel that a new legal order must be established.

Those unacquainted with events and highlights of the past decade's effort to forge a new law may wonder why there is such a delay in reaching an accord. What this chapter attempts to illustrate is that the mining of manganese nodules has been approached by the international community as something other than a narrow and traditional mining enterprise. A 1970 U.N. resolution, called the Declaration of Principles, declared that the manganese nodules found on the seabed beyond national jurisdictions were to be the "common heritage of mankind." The introduction of this new legal and political standard has fundamentally altered the terms of reference and complicated the search for an appropriate accord.

The division of ocean resources has, until relatively recently, been an issue of low salience on the international agenda, but certainly not because the oceans—which cover approximately three-quarters of the globe—were perceived to be deficient in resources. Instead, the reasons for its low political significance were that man lacked the capability to exploit these potential resources intensively, thereby eliminating cause for political conflict, and that most exploitation that did occur was confined to coastal regions where sea law was established and generally respected. If conflict did arise, therefore, it was among regional neighbors and did not assume global proportions. There never has been a serious conflict over ocean resources directly involving the superpowers.

There is no question, however, that we have opened up a new era of seabed exploitation. This era may herald not only greater exploitation but also increasing disputes over the division of the spoils. UNCLOS has been convened, therefore, in order to forestall such disputes and to prevent clashes of force.

The motif of this new era is, of course, the extension of man's capabilities, or in other words, technology. We use the oceans today for essentially the same purposes as previous generations. What advancing technology has changed, however, is the scale and intensity of our utilization. Where once ocean-going vessels lobbed cannonballs to produce destruction, submarines now lurk underwater with weapons capable of destroying entire civilizations. Where once sleek sailing ships carried spices and gold from one continent to another, now superships of more than 200,000 dead-weight tons carry a cargo of "black gold" to societies desperately dependent upon it. Moreover, the line and bait of the coastal fisherman have in many areas been replaced by the nets of huge trawlers and accompanying factory, or processing, ships.

These and other changes in the scale and intensity of ocean use have profoundly altered our perception of the appropriate legal and political foundations for ocean exploitation and intercourse. When man's use of the oceans was essentially limited, there appeared little reason for an explicit and detailed regulatory framework. The Dutch jurist Grotius in the seventeenth century set forth the basic rationale for "freedom of the seas," a principle which was to prevail for centuries: "The sea can in no way become the private property of any one, because nature not only allows but enjoins its common use. . . . Nature does not give a right to anybody to appropriate such things as may inoffensively be used by everybody and are inexhaustible, and therefore, sufficient for all."[1] It was the vision of a limitless and inexhaustible sea that shaped the basic legal framework before World War II and argued against the extension of national sovereignty over resources.

Technology developed after World War II, however, has led to the erosion of this vision and the consequent applicability of the traditional legal framework. There is currently no uniform international standard delineating national jurisdiction over ocean space and resources. At last count, forty-two nations had made territorial sea claims of from 3 to 10 nautical miles; fifty-four nations claimed territorial seas of 12 nautical miles; and twenty nations claimed territorial seas beyond 12 nautical miles, nine of them claiming as much as 200 nautical miles. National claims over ocean space for zones of special purpose, such as fishing, demonstrate equal diversity.[2]

Currently, therefore, there is agreement among the international community neither on where national jurisdiction over ocean space ends nor on what legal regime exists beyond national boundaries. Virtually the only principle on which there is international consensus is that nations have no sovereign territorial rights to any part of the deep seabed or deep oceans. In other words, it can be said with certainty that there will be no national territorial claims to an undefined portion of the deep seabed that will be legitimized by the international community.

Regimes

Lawyers have argued for quite some time about what constitutes the current legal regime for the deep seabed. Historically, most lawyers have claimed that the resources of the seabed beyond national jurisdiction are *res nullius,* i.e., these resources do not belong to anyone but are the property of the first party that can possess them. Hence, they are open to acquisition on a first-come, first-serve basis. On the other hand, some lawyers have claimed that the deep seabed is *res communis,* meaning that instead of belonging to no one, the deep seabed belongs to all. In this interpretation, the resources of the seabed are subject to the inclusive use of all nations and would not be subject to any state's appropriation or sovereignty. Thus, this principle would not necessarily permit exploitation to take place simply because a party has the technology in hand. Developing nations, as a rule, prefer the *res communis* interpretation since it would eliminate the possibility of mining companies gathering nodules of their own volition.

For the most part, this debate among lawyers has been academic. As we will see, whereas the U.S. government prefers to accept the *res nullius* principle, it is not a sufficient basis upon which to launch an enterprise. Mining companies have expressed a desire to have far more security than is afforded by an abstract legal doctrine. For the reasons mentioned in Chapter One, a set of detailed rules and regulations to govern mining in the deep seabed is required to provide the necessary climate for investment.

A landmark in delineating where national seabed ends and international seabed begins was the Truman Proclamation of 1945, which unilaterally extended jurisdiction by the United States over the natural resources of the continental shelf contiguous to the mainland. Hollick has stated that this proclamation "sought to facilitate conservation of shelf resources, to provide protection against foreign exploitation of those resources, and to promote domestic investment in offshore mining by assuring U.S. industry security of tenure."[3] To be sure, the Truman Proclamation was not the first national claim to seabed resources beyond territorial limits in the twentieth century, as other nations had made such claims.[4] The fact that the supreme global power of the period had taken such action, however, lent a legitimacy and impact to the claim that went well beyond previous ones. This unilateral extension of national jurisdiction initiated a chain of reciprocal unilateral claims, which ultimately led several Latin American nations to claim total national jurisdiction over both the seabed and superjacent waters 200 miles from their coasts.

The desire to establish uniform offshore claims drew legal scholars in the early 1950s to convene for the purpose of producing a definitive legal standard or convention that statesmen from all nations could subsequently ratify. This gathering resulted in the second milestone in seabed policy, the 1958 Convention on the Continental Shelf. According to this convention, coastal states were granted exclusive jurisdiction over the resources of their contiguous continental shelf—an extension, internationally, of the Truman Proclamation. The major problem with this seemingly unequivocal standard was that not all continental shelves extend, or slope, from the land mass equally; hence, the problem of uniformity remained. The authors of the 1958 convention responded by defining the continental shelf as follows: "The seabed and subsoil of the submarine areas adjacent to the coast . . . to a depth of 200 meters or beyond that limit, to where the depth of the superjacent waters admits of the exploitation of the natural resources of the said area."[5]

This definition was not so much based upon the geophysical features that characterize the continental shelf as it was upon the state of technology to exploit seabed resources. The drafters of the 1958 Continental Shelf Convention did not perceive how obsolete their legal standard would become in a relatively short period of time. Less than five years later, mining companies were seriously contemplating the mining of manganese nodules and beginning the task of prospecting. Once it was clear that seabed mining could in the not-too-distant future extend far beyond the 200-meter isobath, a dispute arose as to how literally the 1958 convention should be taken. Some jurists, such as Japan's Shigeru Oda, claimed that, in effect, it carved up the seabed on a national basis, permitting no seabed commons: "It can be inferred that, under this Convention, all the submarine areas of the world have been theoretically divided among the coastal states at the deepest trenches. This is the logical conclusion to be drawn from the provision approved at the Geneva Conference."[6] Other jurists contended that any conclusion stating that the seabed was made up of nothing but continental shelves was absurd. Indeed, would the seabed belong exclusively to those nations that possessed the capabilities to mine it? This interpretation was not likely to find favor in the international community. Some have claimed that the language of the convention indicates a clear restriction over national claims to the seabed by explicitly dealing with "The seabed and subsoil of the submarine areas *adjacent* to the coast." It could be interpreted, therefore, as only dealing with the seabed that is close or proximate to the coast, thereby eliminating the deep seabed from national appropriation.

There is little doubt that the drafters of the convention did not intend to set forth a basis for national division of the entire seabed. They assumed that once technology reached out to the deepest area of the oceans the international community would again be convened to provide a new basis for legal intercourse; but they had no idea the time would come so soon.

There is another objection to the 1958 convention often raised by third-world international scholars; that is, the convention was only ratified by approximately forty nations. Since in 1958 there were far fewer sovereign states than now exist, there is a good deal of skepticism about the legitimacy of a legal doctrine ratified by what now is only a part of the international community. Nations that have gained statehood since 1958 often reject as binding not only the 1958 Convention on the Continental Shelf but other conventions from the period as well.

Some of those who reject the 1958 convention's applicability claim that the deep seabed is currently subject to traditional high-seas freedoms; that is, the resources thereon would be *res nullius* and subject to appropriation by the party first capturing them. Hence, freedom of exploitation would be allowed, subject to respecting and observing the right of others to engage in similar activities. Unlike the 1958 convention, this legal doctrine does not bestow the exclusive right of exploitation to any party. It therefore posits that vast areas of the seabed are not subject to national jurisdiction but open to any nation that can effectively appropriate the resources.

As stated previously, when there was no anticipated activity on the deep seabed, the dispute over its legal status was merely academic. Mero, as early as the late 1950s, recognized that the prospect of exploitation would bring legal questions to the forefront. In 1960 he wrote, "Mining of nodules from the deep sea may be attended by legal complications, because marine law is quite vague. It has been kept so by the implied mutual consent of the nations of the world. . . . A more rigid law should probably be written defining the line on the ocean floor at which national boundaries end."[7] In his book published in 1965, he again expressed concern over the legal ambiguities that existed and prophetically ventured that it might take an extended period of time to draft a new legal regime. In what appeared more hope than conviction, he claimed that universal acceptance of the *res nullius* interpretation would benefit all concerned.[8] Obviously, he did not anticipate the objection developing nations would have to this legal principle or the negative reaction of the mining companies themselves.

Arvid Pardo

Mero's key role was that of a mining pioneer and entrepreneur; but the role of legal catalyst fell on the shoulders of another: the Maltese Ambassador to the United Nations, Arvid Pardo. Pardo, on August 17, 1967, submitted a request for the inclusion of a supplementary item on the agenda of the twenty-second session of the General Assembly. The initiative requested the General Assembly to consider the formulation of a treaty declaring the seabed and ocean floor beyond national jurisdiction the "common heritage of mankind." The request was accepted, and on November 1, 1967, Ambassador Pardo spoke eloquently for over three hours on the necessity of formulating a new international regime. The focus of his address was that technology to exploit the oceans was advancing so quickly, and the legal framework for exploitation was so ill-defined, that these factors would lead to conflicting claims of national jurisdiction over the seabed unless the international community acted immediately to resolve the issue. Painting a picture of the vast wealth of ocean resources, he stated, "It is clear that the seabed beyond the 200-meter isobath will soon be subject to exploitation. The only question is, would it be exploited under national auspices for national purposes, or would it be exploited under international auspices and for the benefit of mankind? The wording of the 1958 Convention, whatever may have been the intentions of its authors, provides powerful legal encouragement to the political, economic and military considerations that are inexorably impelling technologically advanced states to appropriate the seabed and the ocean floor beyond the 200-meter isobath for their own use."[9]

Pardo feared the force of technological developments leading inextricably to national jurisdiction over vast stretches of the seabed, just as it had in the earlier Truman Proclamation. The division of the seabed among the industrial nations could have been legitimized under the guise of the 1958 convention. Pardo was also concerned about possible heightened military activity. In short, he feared a competitive scramble for resources that would surpass in magnitude former colonial territorial acquisitions. Unless direct intervention of the entire international community was forthcoming in the creation of a new legal regime, he stated, mankind would later regret the plunder and conflict over ocean resources that would otherwise result. His preferred solution to the anticipated problems was the creation of an international agency or organization which would have jurisdiction over resource exploitation on the seabed beyond a clearly delineated national boundary. He called upon the General Assembly to form a group that would declare the unclaimed seabed the "common heritage of mankind" and would prepare for the establishment of an appropriate international body.

It would be a mistake to leave the impression that Pardo's initiative was totally original. The idea that some form of international compact should regulate activities beyond national jurisdiction–rather than allowing the dominance of a *res nullius* legal standard–goes back to at least the previous century. There were occasional statements made to this effect during the 1958 and 1960 Geneva Law of the Sea Conferences, but nothing came of them. During the mid-1960s, however, a resurgence of interest in internationalization arose. In 1965, the Commission to Study the Organization of Peace called for vesting title to the entire ocean area beyond the 12-mile territorial limit (for living resources) and the continental shelf (for minerals) with the United Nations. In 1967, the World Peace Through Law Conference adopted a resolution calling for the resources of the oceans beyond the continental shelf to be placed within U.N. jurisdiction. In that year Senator Frank Church called for a similar legal thrust. It should be noted that the purpose of these initiatives was as much to procure a source of independent financing for the United Nations as to bring world order to the oceans. Pardo, recognizing strong objections to this financing, made no such connection in his speech. Although he emphasized that the resources of the deep seabed should, to a large extent, benefit the people in poorer nations, his proposal was purely ocean-directed; it was seen primarily as a means of preventing the outbreak of conflict and only secondarily as a means of distributing wealth.

The Seabed Committee

Although Pardo was not the first to call for international control of the deep seabed, his U.N. position was crucial to the advancement of his plea. It struck a responsive chord among his colleagues, though, as we will see later, it did not gain a uniformly enthusiastic reception. To study the Pardo thesis and examine possible legal regimes, the U.N. General Assembly in 1968 created an Ad Hoc Committee to Study the Peaceful Uses of the Seabed and Ocean Floor Beyond the Limits of Jurisdiction (Seabed Committee). Thus, the United Nations was tackling local seabed issues in a manner similar to its approach regarding outer space nearly a decade earlier. Within a year, the Ad Hoc Seabed Committee was transformed into a more permanent Seabed Committee. The early years were essentially educational in nature, as most U.N. delegates had no more than the vaguest ideas about the dynamics of ocean resources and the legal complexities surrounding them. The committee expanded in membership almost yearly, and therefore the educational process had to be repeated constantly for incoming members. Unlike the preparations for the first Law of the Sea Conference held in 1958 in Geneva, there was no independent technical or legal body called forth to set the terms of reference or to provide legal guidance. Instead, the U.N. delegates or diplomats assigned were expected to develop their own expertise in this area. Many delegates did so, which later put them in key positions to influence the negotiations that were to take place in the 1970s.

In the early years a number of U.N. resolutions were passed that emanated from the Seabed Committee. Not all the resolutions were of equal importance, but two of the most important were the following:

1. the so-called Moratorium Resolution (2574-D), passed in December 1969, pledging states to refrain from (*a*) resource activities in the seabed beyond national jurisdiction pending the establishment of an international regime, and (*b*) claims of exclusive jurisdiction over seabed resources beyond the limits of national jurisdiction
2. the Declaration of Principles Governing the Ocean Floor, and the Subsoil Thereof, Beyond the Limits of National Jurisdiction, passed on December 17, 1970, establishing the analytical framework within which negotiations were to proceed

The Moratorium Resolution was particularly controversial, as it ran directly counter to the *res nullius* principle espoused by many industrial nations. The vote on this resolution before the General Assembly was fifty-eight for, twenty-nine against, and thirty-five abstentions. It looked for a time as if Summa Corporation was going to mine in direct opposition to the resolution; had it gone ahead, this violation would have raised a storm of international protest.

The practical impact of the resolution, over time, has been debated. In 1969, when it was passed, there was no clear demarcation separating national from international seabed boundaries. It was impossible to determine, therefore, where mining could proceed and where it could not. Industrial nations claimed that the resolution would lead to extensive claims of national jurisdiction to ensure that resources be captured by interested states. However, few extended claims of national jurisdiction since 1969 can be directly traced to the resolution. Since Summa Corporation only used mining development as a cover, and other, more legitimate, mining companies require more than the *res nullius* principle to stake their claim, there has yet to be a violation. Should the U.S. government or any other government unilaterally license miners to exploit deep seabed resources, however, this would be in direct violation of the resolution.

Even more fundamental was the passage in 1970, of the Declaration of Principles, a document that nations struggled two years to produce in the Seabed Committee. Its purpose, as noted in the title, was to set forth the principles upon which a new and definitive legal regime could be based. This task was by no means easy, since nations held radically opposed views about what constituted an appropriate regime.[10] Some nations wanted a statement declaring narrow national jurisdictions, whereas others, most notably the Latin Americans, insisted upon extended national jurisdiction over ocean space. Many developing nations wanted the elaboration of a strong international organization written into the declaration. The socialist bloc, on the other hand, did not want to even concede that an international agency need be created. What resulted from this strongly contested issue was a declaration that constituted an agreement of the lowest common denominator. Nevertheless, it did establish some extremely important principles. Article 1 stated,

> The seabed and ocean floor, and the subsoil thereof, beyond the limits of national jurisdiction (hereinafter referred to as the area), as well as the resources of the area, are the common heritage of mankind.

Article 4 continues,

> All activities regarding the exploration and exploitation of the resources of the area and other related activities shall be governed by the international regime to be established.

Finally, Article 9 stated,

> On the basis of the principles of this Declaration, an international regime applying to the area and its resources and including appropriate machinery to give effect to its provisions shall be established by an international treaty of a universal character, generally agreed upon. The regime shall, *inter alia,* provide for the orderly and safe development and rational management of the area and its resources and for expanding opportunities in the use thereof and ensure the equitable sharing by States in the benefits derived therefrom, taking into particular consideration the interests and needs of the developing countries, whether land-locked or coastal.

The Declaration of Principles is notable for what it says and what it does not say. Unable to resolve the question of separating national from international jurisdictions, the declaration simply leaves the issue open. In other words, it speaks of national and international areas but does not delineate their respective dimensions. Article 9 does establish that an international organization or machinery will be created for the management of resource exploitation in the area, but it fails to detail what kind of machinery would be appropriate. The issue of how strong an organization this agency should become, therefore, is left open. By drafting the Declaration of Principles, the international community was not able to resolve the deep schisms evident in Seabed Committee deliberations. What the international community was able to do by forging the declaration was to set the course for negotiations that were to follow—no mean accomplishment.

Unlike the Moratorium Resolution, the Declaration of Principles gained nearly universal support. The vote in the General Assembly was 108 nations for, none against, and fourteen abstentions (coming primarily from the Eastern European bloc which opposed the proposal to create international machinery).

It is true that many nations, including the United States, do not interpret U.N. resolutions as binding international law. The U.S. government has hastened to publicize its view that the legal status of deep seabed resources is not in any way affected by the passage of the Moratorium Resolution. Companies would not, in effect, be prohibited from mining beyond national jurisdiction at this present moment. Many other nations, however, view the two resolutions as marking the beginning of a new era in ocean exploitation. Knight, speaking in particular about the Declaration of Principles, has stated, "Prior to the adoption of the Declaration of Principles, manganese nodules would have been regarded as *res nullius* in the same manner as fish swimming in the high seas. Resolution 2749, however, indicates a contrary expectation on the part of the vast majority of the international community."[11] Indeed, the vast majority of nations now believe that the fundamental basis of exploitation has been changed, and such beliefs weigh strongly in formulating a customary rule of law. Hence, whereas neither resolution settled or resolved the disputes, their importance in forging a counter position to that of the industrial nations should not be underestimated.

Deepsea Ventures

Mining organizations in the United States, impatient with prolonged UNCLOS negotiations and pessimistic about the chances of this forum producing a regime that fosters a stable and favorable investment climate, have urged U.S. government support for what they claim would be interim mining, i.e., sanctioning and regulation of mining before the completion of a definitive UNCLOS treaty. These organizations claim that it is not only in their interest to move ahead quickly but in the national interest as well. Obviously, such a development would violate the two previously noted U.N. resolutions and create a full-blown international issue.

There has been no confrontation as yet because the U.S. government has not seen fit to tender its support for what the industry feels is desirable. A test case came in November 1974, when Deepsea Ventures filed a legal claim for exclusive mining rights to approximately 60,000 square kilometers of Pacific Ocean seabed. In addition, the company requested "diplomatic protection" from the U.S. State Department while conducting the mining. The basis for the claim and the response to it deserve considerable elaboration.

The 60,000-square-kilometer area selected by Deepsea Ventures was precisely the following: From

lat. 15° 44′ N long. 124° 20′ W

west to

lat. 15° 44′ N long. 127° 46′ W

then south to

lat. 14° 16′ N long. 127° 46′ W

and then east to

lat. 14° 16′ N long. 127° 20′ W

The claim to this specific ore body was to extend for forty years, but the company stated that after fifteen years at the site, the area claimed would be reduced in half (to 30,000 square kilometers). The site was estimated to be more than 1,000 kilometers from the nearest island and 1,300 kilometers beyond any nation's claim to the continental shelf or margin. The request for diplomatic protection was "for the purpose of facilitating the protection of Deepsea's rights and investments should this be required as a consequence of any future actions of the United States Government or other States, persons, or organizations."[12] It was felt that since corporations do not possess legal standing to appear before international tribunals, they must rely on states to protect their rights.

Deepsea Ventures claimed that the cost of prospecting, exploration, and research and development associated with the specific site had been approximately $20 million up to that time, and they further claimed that anticipated costs over the next three years would be from $22 to $30 million. From the ore site, Deepsea Ventures anticipated gathering 1.35 million metric tons of nodules per year.

The legal brief for this claim was written by the law offices of Northcutt Ely, Washington, D.C. Ely's major legal point was that Deepsea Ventures was entitled to possess an exclusive claim over deep seabed resources, based on the following:[13]

First, he stated that there were precedents for exclusive claims to deep seabed resources. Ely pointed to a number of national assertions before 1945 over resources found on the seabed clearly beyond national jurisdiction (e.g., England and France had in the past claimed jurisdictions over oysters, Libya and Turkey over sponges, Panama and the Philippines over pearls). Recognition of their rights occurred, he claimed, when they demonstrated "occupation" of the desired area. Occupation, he stated, was equivalent to legal possession. Of course, one cannot occupy areas of the seabed 2 miles or more below the surface of the ocean. Occupation or possession is demonstrated, he stated, by proceeding with "reasonable diligence" in bringing the desired resources into the commercial market. Such diligence, he felt, was adequately demonstrated by Deepsea Ventures' prior research and development, as well as the substantial sums that the company had invested. Moreover, customary international law recognized that exclusive claims could be made if the resources were clearly subject to uninterrupted exploitation. What was required was actual exploitation of the particular resources in question (the 1958 Continental Shelf Convention would then cover the exclusive mining claim, since the convention allowed one to proceed as far as technology would allow).

Moreover, Ely claimed, the 1958 Convention on the High Seas sanctioned manganese nodule mining. This convention did not explicitly list such mining as one of the freedoms allowed beyond the limits of national jurisdiction, but neither did it explicitly exclude it. To bolster the position that nodule mining is a legitimate exercise of freedom of the seas, Ely cited the statements of the State Department's legal advisor, Charles N. Brower:

> At the present time, under international law and the High Seas Convention, it is open to anyone who has the capacity to engage in mining of the deep seabed subject to the proper exercise of high seas rights of other countries involved.[14]

Finally, Ely claimed that there was an absence of any prohibitory rule in traditional international law preventing the exploitation of manganese nodules beyond national jurisdiction. The resolutions passed by the United Nations (cited earlier), he claimed, did not constitute international law, since the U.N. General Assembly is not vested with legislative power and hence its resolutions are not binding upon member states. Ely stated, "A General Assembly resolution is merely an invitation or recommendation to member States to enter into a treaty relationship concerning the matters dealt with in the resolutions."[15] Official U.S. government statements were also cited, since, as mentioned previously, the U.S. government also does not recognize U.N. resolutions as legal and binding.

Ely summed up his argument by saying, "The practice of States, the 1958 Convention on the High Seas, and the General Principles of Law recognized by civilized nations, taken together, justify the conclusion that a positive rule of international law permits the acquisition of exclusive rights to deposits of manganese nodules in seabed areas beyond national jurisdictions."[16]

The Deepsea Ventures claim caused a stir among the international community, though it failed to change any positions. Most national delegates simply ignored it, but some nations, such as Canada and Australia, officially informed the U.S. government, as well as Deepsea Ventures, that they did not recognize the validity of the claim. The claim was attacked directly the next year in an article in *International Lawyer* written by Gonzalo Biggs, a lawyer with the Interamerican Development Bank.[17]

Biggs states that the cases regarding exclusive jurisdiction over sponges and oysters cited by Ely involved rights granted to coastal *states* and not to the first discoverer of these resources. Deepsea Ventures, as a corporation, then, could not in any way be accorded exclusive mining rights. Biggs also brought out the contradiction between claiming that resources found beyond the 200-meter isobath could be subject to exclusive jurisdiction (as implied by the 1958 Continental Shelf Doctrine), while at the same time, mining could be guaranteed as a high-seas freedom (as Ely claimed was guaranteed by the 1958 High Seas Convention).[18] Biggs, in contrast to Ely, placed a good deal of legal importance upon the 1969 and 1970 U.N. resolutions, stating that they indicated sufficient evidence of general international consensus to form the basis of customary law.

But it was not an academic debate on legal principles that was to resolve the issue. Instead, it was the response of the U.S. government, i.e., the State Department, that put to rest any hopes of Deepsea Ventures for immediate mining. In a letter to the company, the department wrote that it does not grant or recognize exclusive mining rights to mineral resources on the seabed that are beyond the limits of national jurisdiction. Rather, the department considered mining in this area to be a high-seas freedom. Furthermore, it cited UNCLOS, not the unilateral claims of mining companies, as the proper forum for determining international law. International reaction to the claim was muted since the U.S. government itself delivered the rebuff that led Deepsea Ventures to pursue the matter no further. Whatever hopes the company had of developing a functioning international law were dashed by the State Department's refusal to provide diplomatic protection.

U.S. Position

The position the U.S. government has adopted regarding the current legal status of the area beyond national jurisdiction is intriguing since it virtually satisfies no one. Neither U.S. corporations nor most other nations support the government's interpretation of existing law, for differing reasons.

As has been stated previously, the policy that deep seabed mining is a high-seas freedom does not satisfy the requirements of seabed miners. Mining company irritation with the U.S. position has been evident during the many years that have gone into forging a new, definitive legal basis for ocean interaction. Particularly upsetting to mining executives have been repeated statements by high administration officials that companies are currently free to mine as they please beyond national boundaries, when these officials are quite aware that such statements do not provide the guarantees of exclusive rights essential to any land-based or ocean-based mining. Speaking before a Senate subcommittee in 1975, Under-Secretary of State for Security Assistance Carlyle Maw proclaimed, "We encourage private investors to develop their technology and to begin mining when they are ready."[19] Dubs of Kennecott Copper in subsequent Congressional testimony referred to Maw's statement as follows:

> This like other motherhood statements is a fine sentiment. However, it provides zero assurances as to the future financial rules under which a miner would operate or as to whether he would be permitted to mine his hard chosen spot on the seabed under some distant resolution of the law of the sea treaty now being negotiated. Furthermore, the miner has no protection against the losses which would occur if his mining operations were interfered with under the present high seas regime. It is simply not feasible to invest under such circumstances and furthermore, it is not feasible to obtain funds to invest even if the entrepreneur were a high rolling risk taker.[20]

Mining officials have made it clear that the nodule miner must have certain rights in order to ensure the large investment and the integrity of the entire operation. One mining official has identified four minimum requirements: the right to explore for desired minerals, exploit an exclusive deposit, have continued access to the chosen deposit, and have space for ancillary purposes.[21] The U.S. high-seas policy provides for none of these rights. Moreover, as Dubs has pointed out above, prospective nodule miners cannot even obtain the risk capital from banks on the basis of such a policy.

The Common Heritage of Mankind

The U.S. position has not found favor with most other countries either, notably the developing nations. By blithely ignoring the substance of 1969 and 1970 U.N. resolutions, and contending that high-seas freedom still prevails with regard to the seabed, the U.S. position runs directly counter to that of developing nations. Although it continues to give verbal adherence to the principle of the common heritage of mankind, when the United States is pressed it appears to the rest of the world that this support is directly refuted by more forceful adherence to the high-seas or *res nullius* principles. In other words, there seems to be a clear contradiction between the two.

The United States position on the common heritage of mankind is indeed quite curious and deserves brief review. In 1966, President Lyndon Johnson made a bold statement which was to be echoed time and again during subsequent debates. "Under no circumstances must we even allow the prospects of rich harvest and mineral wealth to create a new form of colonial competition among the maritime nations. We must be careful to avoid a race to grab and hold the lands under the high seas. We must ensure that the deep seas and the ocean bottom are, and remain, the legacy of all human beings."[22] Pardo's statement in 1967 about the "common heritage of mankind" so closely paralleled Johnson's "legacy of all human beings" that the United States could hardly repudiate the term. What U.S. officials have claimed, and continue to claim, is that although they can support the term, the common heritage of mankind, for the deep seabed, it implies no explicit or agreed upon legal principle. Hence, from this point of view there is no contradiction between the legal principle of *res nullius* and the essentially *moral* principle of the common heritage of mankind. It is clear that President Johnson in 1966 was invoking only a moral obligation which the international community had with respect to the deep seabed and its mineral wealth. Foreign delegates read more into the phrase than do U.S. negotiators, and the position of the latter has remained firm. In its most basic sense, U.S. officials have most often interpreted the injunction of common heritage to mean there ought to be some sharing of the economic rent derived from mining with the entire international community, particularly with the developing nations.

In general, one seldom finds any mention of the common heritage in any of the speeches or pronouncements of U.S. officials—no doubt because of the more expanded meaning brought to the phrase by developing nations. The United States firmly denies that the phrase has any specific legal meaning, as the chief negotiator on the seabed issue for many years stated in congressional testimony: "In essence, we defined our understanding of the term common heritage to mean whatever the collection of treaty articles ultimately means, that the term would have no independent meaning. . . . We think that the common heritage of mankind principle, if I can be blunt, is a glittering generality. Therefore, we can leave the phrase alone, provided the rest of the treaty says the right things."[23]

To the developing nations, however, the fact that deep seabed minerals are the common heritage of mankind represents more than simply a slogan. There are common legal interpretations of the phrase running through the usage of many national delegates and independent observers. First, in its most fundamental sense, the common heritage of mankind is meant to refer to property rights. Unlike the United States, many nations claim that it implies common ownership of resources or common property.[24] Second, it is interpreted as meaning that this common property cannot be appropriated without first obtaining the consent of states, either through a treaty or establishment of an appropriate international organization.[25] If these are the elements upon which this principle is predicated, it follows that any area or resources covered by it cannot be appropriated without the express approval of the international community. In other words, those nations possessing the technology to exploit these resources are prohibited from doing so until the international community or its legitimate representative establishes the terms of exploitation.

Even more controversial than the legal has been the political interpretation of the phrase. Kent claims that the following moral and political imperatives have become a part of what constitutes the common heritage: (1) the resource should only be used for peaceful purposes; (2) the benefits derived from the use of the resource should be distributed equitably; (3) users of the resource should show respect for the environment, limiting both depletion and pollution; (4) a governing agency should be formed to ensure the implementation of the three previous principles, and this agency should provide for participation by all affected parties in making its decisions.[26]

It is particularly the call for full international participation in resource management that is heard today. The Venezuelan representative to UNCLOS in 1973 stated,

> To developing nations the concept of common heritage implies not only sharing in the benefits to be obtained from the exploitation of the resources of the area, but also, and above all, an effective and total participation in all aspects of the management of this common heritage.[27]

In reality, "participation" is too mild a word to describe the goal of developing nations as reflected in their efforts at the various UNCLOS sessions. What in fact has been the goal of the G-77 (the negotiating group for developing nations) since 1974 has been total and effective *control* over area activities, not simply participation.

To most developing nations, therefore, the common heritage of mankind represents a combination of moral, legal and political principles, which is certainly not what the United States was voting for when it supported the 1970 Declaration of Principles. Some have noted, nevertheless, that should the United States at some time in the future unilaterally license domestic companies to mine the seabed, or simply allow miners to exploit nodules as a high-seas freedom, the U.S. vote in favor of the Declaration of Principles could come back to haunt it, i.e., in a court case before the International Court of Justice. Realizing this dilemma, Northcutt Ely wryly noted in congressional testimony, "We should apparently regard the unfortunate vote of the American representative for the 1970 Resolution as being due to a bad telephone connection with Washington."[28]

Conclusion

The difficulties with the U.S. legal position are plain to see. Mining companies are quick to point out that the principle of high-seas freedom, in practice, allows no freedom at all, since there is no security of tenure. Third-world nations, on the other hand, note that sanctioning the mining of nodules under this principle, while at the same time proclaiming allegiance to that of the common heritage of mankind, represents a contradiction. However, since there has been no commercial mining entry as yet, this contradiction has remained dormant. There would be little concern over present policy among mining officials if they were convinced that U.S. negotiators would be able or willing to see their needs met in the formation of a new regime. However, strident and persistent support by developing nations for the application of the principle of the common heritage of mankind, and all that it entails, raises serious doubt as to whether U.S. negotiators can fulfill the perceived needs and desires of the nodule industry.

Notes

1. H. Gary Knight, *The Law of the Sea: Cases, Documents and Readings* Washington, D.C.: Nautilus Press, 1975).
2. R. W. Smith and R. D. Hodgson, "Unilateralism: The Wave of the Future," Paper presented at the 11th Annual Conference of the Law of the Sea Institute, Kingston, R.I., June 1976.
3. Ann L. Hollick, "U.S. Oceans Policy: The Truman Proclamations," *Virginia Journal of International Law,* 17, 1 (Fall 1976), 24.
4. See Northcutt Ely, "International Law Applicable to Deepsea Mining," in *Current Developments in Deep Seabed Mining,* Hearings before the Senate Subcommittee on Minerals, Materials, and Fuels, 94th Congress, 1st Session, November 7, 1975.
5. H. Gary Knight, "Issues Before the Third United Nations Conference on the Law of the Sea," *Louisiana Law Review,* 34 (1974), 159.
6. As quoted by Gaetano Arangio-Ruiz, "Reflections on the Present and Future Regime of the Seabed of the Oceans," Symposium on the International Regime of the Seabed, Rome, June/July 1969, p. 2.
7. John L. Mero, "Minerals on the Ocean Floor," *Scientific American* (Dec. 1960), p. 71.
8. John L. Mero, *The Mineral Resources of the Sea* (Amsterdam: Elsevier 1965), p. 292.
9. Arvid Pardo, in *Interim Report on the UN and the Issue of Deep Ocean Resources,* House Committee on Foreign Affairs, 90th Congress, 1st Session Dec. 7, 1967, p. 277.
10. Johan Ludvik Løvald, "In Search of an Ocean Regime: The Negotiations in the General Assembly's Seabed Committee 1968–1970," *International Organization,* 29, 3 (Summer 1975), 681–709.
11. Knight, "Issues Before the Third United Nations Conference," p. 170.
12. Deepsea Ventures, "Notice of Discovery and Claim of Exclusive Mining Rights, and Request for Diplomatic Protection and Protection of Investment, by Deepsea Ventures, Inc." in *Current Developments in Deep Seabed Mining,* p. 36.
13. From Opinion of the Law Offices of Northcutt Ely, "International Law Applicable to Deepsea Mining," in *Current Developments in Deep Seabed Mining,* pp. 62–127.
14. Hearings before the House Subcommittee on Oceanography of the Committee on Merchant Marine and Fisheries, 93rd Congress, 1st Session, 1974, as cited in Ely, "International Law," in *Current Developments in Deep Seabed Mining,* pp. 165–66.
15. Ely, "International Law," in *Current Developments in Deep Seabed Mining,* p. 121.
16. *Ibid.,* p. 125.
17. Gonzalo Biggs, "Deepsea's Adventures: Grotius Revisited," *International Lawyer,* 9, 2 (1975), 271–81.

18. *Ibid.*, p. 278.
19. As cited in *Current Developments in Deep Seabed Mining,* p. 4.
20. *Ibid.*
21. C. H. Burgess, "Needs of the Mineral Industry," in *The Law of the Sea: International Rules and Organization for the Sea,* Proceedings of the 3rd Annual Conference of the Law of the Sea Institute, June 24–27, 1968, p. 327.
22. Address given by President Johnson at the commissioning of the ship *Oceanographer,* July 13, 1966.
23. Statement of Leigh Ratiner in *Deep Seabed Mining,* Hearings before the House Subcommittee on Oceanography, 94th Congress, 2nd Session, Feb. 1976, p. 130.
24. Biggs, "Deepsea's Adventures," p. 280.
25. R. P. Arnold, "The Common Heritage of Mankind as a Legal Concept," *International Lawyer,* 9, 1 (1975), 153–58.
26. George Kent, "The Common Heritage Idea," Unpublished paper (May 1975).
27. A. Aguilar, "How Will the Future Deep Seabed Regime Be Organized," in J. K. Gamble and G. Pontecorvo, eds., *Law of the Sea: The Emerging Regime of the Oceans* (Cambridge, Mass.: Ballinger, 1974), p. 47.
28. Ely, "International Law," in *Current Developments in Deep Seabed Mining,* p. 177.

Chapter 3

The Impact of Seabed Mining

When the technical feasibility of mining manganese nodules first became generally recognized, the prospect for commercial exploitation was viewed with considerable excitement. The possibility of tapping this abundant ocean resource raised expectations in some quarters that the entire international community could immediately reap the benefits. The past decade of deliberations and analysis, however, have brought lower expectations and great disillusionment. This chapter discusses the dispute prevalent in recent years over who, in fact, will benefit from manganese nodule mining and who might be negatively affected.

This chapter also focuses upon efforts by U.S. representatives to convince the international community that manganese nodule mining, free from regulations affecting price or levels of production, will benefit all mankind. As will be seen, those nations that currently export the minerals found in manganese nodules have, for understandable reasons, viewed such mining as something other than a desirable development. What is less obvious, and what constitutes the major contribution of this chapter, is the analysis of why many other nations—quite removed from export considerations—have remained unconvinced of the U.S. thesis.

The benefits from mining, in general, often appear self-evident. The struggle of miners to win exceedingly small concentrations of the desired minerals from the foundation in which they are embedded is certainly not elegant but, nevertheless, is required to sustain the functioning of an industrial society. The desired minerals found in manganese nodules are widely used and required in an industrial society; but to assume that other nations, particularly the less industrial ones, will find such mining as compelling and urgent as does the United States is to ignore the very real differences among nations. The case that U.S. negotiators in UNCLOS proceedings have attempted to make is that (1) manganese nodule mining is a universally beneficial enterprise, and (2) these universal benefits can only come about by meeting the perceived requirements of nodule miners.

The inability of U.S. negotiators to make this a persuasive case goes to the heart of the controversy. In fact, the arguments marshalled by the negotiators have been so transparently self-serving that the United States never has gained the credibility required to convince others of their merits. As we will see in this and forthcoming chapters, the U.S. team, in one sense, never grasped the seriousness of differing economic perspectives and thus never made a serious attempt to reconcile the different interests of mineral-consuming nations. In another sense, U.S. officials took the issue too seriously, i.e., the importance in the negotiations of developing nations that are land-based mineral producers.

The potential global impact of manganese nodule mining has been at the heart of the dispute over mining for more than a decade now.

In 1975, the Massachusetts Institute of Technology (MIT) brought together a diverse group of experienced individuals in the field to discuss this issue. The panel consisted of a U.S. government representative, an academician, two mining corporation representatives, and a government representative from a developing nation. Though presumably speaking on the same subject, the respective speeches bore little resemblance to one another.[1] Jack Flipse of Deepsea Ventures spoke of the engineering feat required to raise nodules and what he felt was the lack of government support. Marne Dubs, his business competitor from Kennecott Copper, spoke of the resources in the ocean and industry requirements for their mining. Leigh Ratiner, of the U.S. Department of the Interior and chief U.S. negotiator on manganese nodules at UNCLOS, spoke of an international treaty that would satisfy both major mining companies and the resource needs of the nation. Roger Burns of MIT spoke of the scientific nature of manganese nodules; and finally, Sergio Thompson-Flores of Brazil discussed the need to create a strong international organization to ensure that benefits from mining would accrue to all nations. Perhaps the most striking divergence of opinion was that of Thompson-Flores:

> The argument was made by Mr. Ratiner, and repeated extensively by him last year in Caracas, and the same argument has been presented today by Mr. Flipse, that in fact everybody would benefit if those who are technologically capable could exploit the area, and if the products of this exploitation could be distributed worldwide. But, to those who have studied economics and who have dealt with international negotiation in the economic field, this argument is entirely without foundation. The benefits of any exploitation accrue mainly to those who undertake the exploitation, and seldom to others.[2]

This statement was later sharply rebutted by Dubs: "One comment Thompson-Flores made, was that the benefits of a resource accrue mainly to those who undertake exploitation. I think that this is a fallacy that has been indulged in by the developing world too long."[3]

Unfortunately, the argument, which raised fundamental differences of opinion, was not pursued by the panel. Developing nations perceive large countries and large mining companies as the primary if not exclusive beneficiaries of mining since they possess both the technology to mine and the greatest need for the minerals. U.S. officials, on the other hand, claim that in the long run, all mankind will benefit from the mining of manganese nodules.

Because the question of who benefits from mining has been on center stage at

UNCLOS, we have had, in effect, a particularly good example of technology assessment by the international community. The assessment has not tried to peer too far into the future, not venturing beyond five to ten years after the conclusion of an UNCLOS treaty. The short-term focus is both understandable and proper, given the major uncertainties that face the forecaster, such as technological development and the economic competitiveness of deep seabed mining. It is not necessary, however, to deal in long-term speculation to raise controversy.

Perhaps the most basic fact shaping the economic impact of mining is that the ratio of the desired minerals in a manganese nodule does not match the level of world consumption. It should be recalled that the probable ratio of minerals in a minimally commercial manganese nodule would be the following: 24 percent manganese, nearly 1 percent nickel, .5 to 1 percent copper, .35 percent cobalt. Total 1970 U.S. consumption of these minerals indicates the extent of their use in an industrial society:

MINERAL	SHORT TONS CONSUMED
cobalt	6,500
nickel	175,000
manganese	1,000,000
copper	2,000,000

As can be seen, there is a wide difference in the use of the minerals in question: e.g., copper consumption exceeds cobalt consumption by a factor of over 300. The raising and processing of from 1 to 4 million tons of nodules per year—the initial varying targets set by the mining enterprises—will have a diverse effect upon the respective mineral markets. Projections of the potential economic impact of manganese nodule mining for the year 1985 were provided in a 1974 report of the U.N. Secretary-General (seen in Table 4).[4] This report, which is probably the most detailed examination of the subject yet attempted, assumed that manganese nodule mining would start in the early 1980s. Now, because of continued delays at UNCLOS and because much technical work remains to be done, the output levels probably will be closer to 1990 than 1985. Whatever the year, Table 4 is based upon six separate mining groups collectively raising 15 million tons of nodules and processing all four minerals.

As can be seen from Table 4, quantities of cobalt raised from mining would be small because of its low concentrations within manganese nodules: only 30 thousand metric tons. Yet because of the small global consumption of cobalt (approximately 60 thousand metric tons, excluding centrally planned economies), the amount raised by only six deep seabed miners could represent fully half of the world production as early as 1985. Another study claims that fully 92 percent of U.S. cobalt imports could be obtained from the seabed by 1985.[5] The flood on the market of this cobalt could have a significant effect upon its market price, lowering it by two-thirds. At such a level, it would approximate the price of nickel; and since nickel and cobalt are substitutes for one another

TABLE 4. The Estimated Impact of Six Nodule Mining Enterprises

	NICKEL	COPPER	COBALT	MANGANESE
Probable Production from Nodules[1]	220	200	30	920
Estimated World Demand (1985)	1,220	14,900	60	16,400
Nodule Production as a Percentage of World Demand	18	1.3	50	6
Estimated Price Decline as a Result of Nodule Mining (%)	0–10	0	66	50 (in specialty markets)

SOURCE: *Economic Implications of Seabed Mineral Development in the International Area: Report of the Secretary General,* (A/Conf. 62/25), United Nations, May 22, 1974, p. 42, and *idem,* (A/Conf. 62/37) Feb. 18, 1975, pp. 5–6.
[1] Thousands of metric tons.

in various applications, the market for cobalt could expand, absorbing at least a portion of the extra supplies.

Manganese represents the opposite case. Because of its large concentrations in the nodule, there could be copious amounts available from deep seabed mining in 1985. Yet manganese is a much more widely used mineral than cobalt, and therefore its market impact would not be nearly as significant. As previously stated, it is questionable whether the pure manganese obtained from nodules will replace the more common ferromanganese used in making steel. For this reason, most mining companies do not plan to separate manganese in processing for marketing. Such large quantities of this material raise the possibility of its use in totally novel applications, but although scientists have been working to discover and produce a large mass market for pure manganese, their efforts have not yet been successful. One should not rule out totally new applications in the long term, however. Oil in the nineteenth century was an abundant material searching for a market, and we are witness to the result today. Deepsea Ventures, as stated previously, plans to market manganese in the small "specialty use" market, and U.N. estimates indicate its price in these markets could fall by as much as 50 percent by the year 1985.

Nickel is the primary mineral of interest in manganese nodules. Some people in the industry have even taken to changing the name to "nickel nodules." The reason for this interest is a combination of high expected growth rates for the use of nickel, high market prices, and an oligopolistic market concentration. As seen in Table 4, approximately 220 thousand metric tons of nickel could be derived from seabed nodules by the year 1985, representing 18 percent of the total global demand. Because of the expected growth of demand, it is not clear whether such an amount would have a discernible effect on market price. However, one study has estimated that because of seabed mining the price of nickel could decline by 10 percent in 1990.[6] Over the long term, therefore, beyond the year 2000, the effect of seabed mining on nickel prices could be substantial.

Copper is an altogether different case. Because of its low concentration in nodules and the large aggregate world demand for it, there is not likely to be any perceptible impact on copper prices by 1985, 1990, or even beyond 2000. The U.N. study estimates nodule copper to capture only a little over 1 percent of the total copper market by 1985. Because of the expected growth in demand, Orr states, "Even with 20 firms producing copper from the ocean in the year 2000 no more than 1.5% of the world's demand for copper could be met from ocean sources. It would take 2,400 firms the size of Deepsea Venture's proposed operation to produce all of the world's copper demand by the year 2000, even under conservative demand projections."[7]

Those uninitiated to the political debate might assume that lower mineral prices would be met with universal applause and that there would be widespread encouragement for the enterprise. Nothing could be further from the case. Instead, the thrust of the debate over the past ten years has been on the effect of seabed mining on existing land-based producers, especially developing nations. Thus attention has been directed toward discovering who are the major producers and exporters of the four minerals in question and the extent of their production.

Table 5 illustrates that a number of developing nations (Group of 77 members) are major mineral producers, mostly of copper, with roles in manganese, cobalt, and nickel, diminishing in that order. The table also illustrates that except for cobalt, most mining for those minerals occurs in major industrial states (60 percent of the world value in these minerals comes from industrial developed nations). Despite the fact that the United States has to import 10 to 20 percent of its copper consumption annually, it is still the world's largest copper producer.

What Table 5 fails to show, however, is the export status of mineral producers, and hence, the potential impact seabed mining would have on specific exporting nations. Unlike the developed nations, developing nations export nearly 100 percent of their mineral production, and such exports invariably represent key export earnings. Industrial nations use their mineral production, on the other hand, for internal consumption. The dependence of developing nations upon a narrow base of exports, and the foreign exchange that is obtained, is well known and needs no elaboration here. Without substantial foreign exchange, developing nations cannot obtain the capital requirements to spur development. Hence, the fact that industrial nations account for the majority of mineral production gives a misleading impression concerning the importance of mining for individual nations. Table 6 shows that there are a number of developing nations

TABLE 5. Approximate 1971 Value of Mineral Production

	COBALT	COPPER	MANGA-NESE	NICKEL	TOTAL	PERCENTAGE OF WORLD OUTPUT
Group of 77 countries	$ 88[1]	$2,602	$ 98	$ 45	$2,333	40.0
Other countries	27	3,523	125	400	4,075	60.0
Total	115	6,125	223	445	6,908	100.0
Group of 77 producers:						
Chile	...	790	...	...	790	11.0
Zambia	10	718	...	...	728	10.0
Zaire	65	449	4	...	518	7.0
Peru	...	235	...	...	235	3.0
Philippines	...	230	...	...	230	3.0
China	...	110	12	...	122	2.0
Mexico	...	70	2	...	72	1.0
Cuba	8	.....	...	27	35	.5
Brazil	...	.....	29	...	29	.4
Gabon	...	.....	20	...	20	.3
India	...	.....	20	...	20	.3
Indonesia	...	.....	...	18	18	.3
Morocco	5	.....	...	...	5	.1
Ghana	...	.....	7	...	7	.1
Other producers:						
U.S.	...	1,522	...	9	1,531	22.0
Canada	11	720	...	186	917	13.0
U.S.S.R.	8	680	76	80	844	12.0
Australia	2	195	11	22	230	3.0
South Africa	...	174	36	9	219	3.0

TABLE 5. continued

	COBALT	COPPER	MANGA-NESE	NICKEL	TOTAL	PERCENTAGE OF WORLD OUTPUT
Japan	...	133	2	...	135	2.0
Poland	...	99	...	...	99	1.0
France	...	.....	...	71	71	1.0
Rhodesia	...	.....	...	9	9	.1
Finland	6	.....	...	...	6	.1
Greece	...	.....	...	9	9	.1

SOURCE: Congressional Research Service, *Ocean Manganese Nodules,* prepared for the Senate Committee on Interior and Insular Affairs, Feb. 1976, p. 45.
[1] Millions of 1971 dollars.

TABLE 6. Mineral Exports and Dependence of Developing Nations (1969)

COUNTRY	EXPORTS (MILLIONS OF DOLLARS)	MINERAL EXPORTS (% OF TOTAL EXPORTS)	MINERAL
Zambia	720.8	94.6	Copper
	4.7	.6	Cobalt
Zaire	475.8	83.0	Copper
	9.1	3.3	Manganese
	29.7	5.2	Cobalt
Chile	730.7	78.3	Copper
Peru	250.0	28.9	Copper
Philippines	150.9	15.6	Copper
Uganda	21.4	10.8	Copper
Haiti	2.3	6.2	Copper
Bolivia	7.4	4.1	Copper
Nicaragua	6.3	4.1	Copper
Mexico	21.5	1.5	Copper
Cuba	13.4	2.1	Nickel
Indonesia	4.4	5.9	Nickel
Gabon	30.1	21.2	Manganese
Ghana	9.1	3.3	Manganese
Brazil	25.8	1.1	Manganese
India	17.6	1.0	Manganese
Morocco	4.4	1.0	Manganese

SOURCE: James C. Orr, *The Economic Effects of Deep Ocean Mineral Mining and the Implications for United States Policy,* Johns Hopkins University, Ocean Policy Project, Occasional Paper Series, No. 4, Dec., 1974, pp. 8–10.

that rely on mineral exports for a significant portion of their foreign exchange earnings. Five countries in particular—Zambia, Zaire, Chile, Peru, and Gabon—show dependencies in excess of 20 percent, and a change in the price of minerals would have major consequences for them. Chile's minister of finance once claimed that each cent decline in the price of copper on the London Metal Exchange means an $18 million loss in Chile's export earnings.[8]

The practical effect of manganese nodule mining on developing nations, however, should not be overemphasized. Although four of the five nations stated above are dependent upon copper earnings for the bulk of their foreign exchange, seabed mining, as we have just pointed out, is unlikely to have a discernible effect on copper prices even beyond the year 2000. The danger to these countries, then, is minimal. In addition, Gabon, the fifth nation with an export dependency in excess of 20 percent, is an exporter of manganese, which may not be extracted in large quantities from the nodules brought to the surface. In short, those developing nations most dependent upon the export of the minerals found in the nodule may not be quite so vulnerable as it appears at first glance.

This fact has not kept the above developing nations, in particular, and all developing nations, in general, from placing enormous emphasis upon the potential negative impact of seabed mining. One reason is that their representatives have not trusted the analyses thus far conducted, which indicate a rather limited effect. To be sure, these analyses have not been carried out upon a "worst-case possible" basis, and therefore do not highlight cases that can be termed "high consequence-low probability." For example, companies might wish to double or even triple what is now considered the probable annual output in order to get a quick return on their capital investment. It has been reported that one company eventually plans on raising 10 to 15 million tons per year, as much as four times the likely beginning output.[9] Should this be the case with all companies, impact estimates would have to be revised substantially.

More fundamentally, how one defines "adverse impact" has never gained consensus among nations. The market price of minerals, which was examined above, is only one standard by which to measure adverse impact. Another is to assess future earnings foregone by projecting what the quantity of exports would be in the absence of deep seabed mining. Regardless of any price impact, seabed mining will reduce the total tonnage of minerals extracted from mines in developing nations. Even if the proportion of seabed mining fails to cover the projected market growth in the future, developing nations can argue that there are earnings foregone. One can also make a case for "investment foregone," namely, the investment capital that would go to seabed rather than land-based mining. Many of the nations listed in Table 6 are developing nations, having high aspirations for the growth of mineral exploitation. Cuba and Indonesia, for example, are relatively recent large-scale nickel producers, and their exports of nickel as a percentage of total exports still only constitute 2.1 percent and 5.9 percent respectively. Their deposits subject to exploitation, however, are quite large, and they view this mining as playing an increasingly important role in their economy. To the extent that investment moves toward the deep seabed, and therefore away from Cuba and Indonesia, it will represent a loss for both countries. Since developing nations as a whole are the least explored areas for mineral deposits on this earth, it has been anticipated that further exploration would be concentrated in the southern hemisphere. The development of seabed technology, however, will retard this movement, at least in terms of the four minerals in question.

Besides the study by the U.N. Secretariat of the impact of deep seabed mining, the United Nations Conference on Trade and Development (UNCTAD) conducted its own independent study.[10] Unlike the former, the UNCTAD report used the concept of "earnings foregone" by developing nations. It claimed that by 1980, or anytime shortly after the initiation of mining, mineral export earnings of developing nations would be $360 million lower than they would be in the absence of seabed mining.[11] The UNCTAD report drew considerable interest at the Caracas UNCLOS session, largely because of the more inclusive attempt to define and identify potential adverse effects.

If the means of measuring adverse impact have been under dispute, so too have been the possible methods of addressing it. The United States and other industrial nations have conceded that some form of compensation should be forthcoming so as not to seriously affect the economies of developing nations. Under a compensatory approach, financial assistance would be provided to those countries most affected, but the size and form of this compensation have never been carefully defined.

The compensatory approach has not found favor among developing nations; e.g., the primary author of the UNCTAD report commented negatively: "Income likely to accrue to the proposed International Seabed Authority would fall short of the potential export earnings foregone by established developing countries. The likely revenue of the Authority would be insufficient to compensate those countries."[12] He also stated that the transfer of automatic lump sums from developed, industrial, consuming nations to make up for a possible shortfall in the authority's funds would be difficult to arrange.[13] Opposition by developing nations to the compensatory approach, therefore, has been based upon the uncertainty of how to calculate the appropriate amount and the fear that neither the international organization nor industrial nations would willingly appropriate sufficient funds.

Developing nations would prefer what is called a "preventive approach," which calls for a prior attempt to eliminate any negative impact or disruption. The development of the seabed industry would be carefully controlled and regulated. Possible production controls include limiting the tonnage of nodules taken, limiting the number of exploitation licenses, or placing "tax brakes" on the amount of nodules exploited. This approach has been strongly opposed by those nations possessing the technology to mine. Before 1976, U.S. representatives claimed that the implementation of production controls can seriously retard necessary investment and unduly delay the gathering of these new and essential resources, thereby denying the benefits of mining to all consuming nations. To place restrictive production controls on the development of an industry before a problem exists, it is claimed, is the wrong way to approach exploitation.

Other means of resolving this issue have been raised (e.g., the formation of comprehensive commodity agreements and a levy on "overabundant" metals, thereby limiting market penetration) but not in detail. Over the many years of negotiation, the fundamental, contrasting approaches, preventive versus compensatory, have remained unresolved; however, as will be seen in Chapter Nine (on negotiating progress), the dispute is now over particulars rather than general philosophy.

The United States' Caracas Initiative

With the studies from UNCTAD and the U.N. Secretariat prepared for the first UNCLOS session in Caracas, there was considerable attention devoted to the effect of mining on land-based mineral producers. The tone of the session, in fact, revolved around how to cope with this seemingly unfortunate intrusion into man's affairs. Little attention was devoted to emphasizing the positive aspects of deep seabed mining. Recognizing this atmosphere, the United States sought to balance the proceedings through an analysis of its own.[14] Besides claiming that the negative impacts upon producers would not be large, this analysis raised several other points worthy of note. First, it posited that consumer interests had been left out of the equation. By mining from the seabed, the study concluded, we would be ensuring that the market prices of the minerals in question would not rise, thereby benefiting the ultimate consumer of goods produced with these metals. The U.S. delegation diplomatically did not raise the point that the elimination of high-price producers is often viewed as a healthy manifestation of a world economy. The report went further by stating that deep seabed mining would benefit consumers, not only within the industrial nations, but also within the developing nations. Finally, it concluded that the imposition of production controls on seabed miners would work to the benefit of large land-based producers—most of whom, as we saw in Table 5, were in industrial nations and not the developing ones. "To the extent there are beneficiaries of restricting seabed production, they will be land-based producers who are largely in the industrially developed countries. Those suffering the greatest losses will be the world's consumers, including the peoples of the lesser developed economies who depend so heavily upon the capital goods made with these minerals for increasing their future standards of living."[15]

Reaction to the analysis was quite negative, partly because of the "hard sell" of the U.S. delegation in general, and the chief U.S. negotiator in particular. Developing nations saw the U.S. initiative for essentially what it was: an attempt to drive a wedge between the developing nations that were consumers and those that were producers. Up to this time developing nations had generally remained united, and U.S. officials assumed correctly that if they could make a split in the united (G-77) position, it would be a significant victory. The problem was, however, that their arguments were based upon ideology (i.e., free-market economics) and hence were unconvincing. No group better knows how skewed the rewards are under a free-market system than the representatives of developing nations. They were hardly prepared to accept liberal economic theory without more empirical confirmation. The U.S. report failed to document how consumers would benefit, when consumers would benefit, and which consumers in what countries would benefit.

U.S. economists, both within and outside the government, have been extremely reluctant to engage in the type of analysis that would illuminate these questions. Those who have addressed the manganese nodule issue inevitably attempt to place it within their traditional economic framework or, if you like, paradigm.[16] The political motives of the U.S. government which led to support for international regulation of mining have led many economists to despair over the "inefficiencies" that might result. Put simply, international regulations are often seen as an unfortunate intervention preventing the efficient and rational functioning of the market. One academic economist, Eckert, has stated, "The literature and policy on regulation of the ocean beds has given insufficient attention to the ways in which regulation could harm economic efficiency by raising the costs of entry into ocean mining, limiting output and introducing monopoly to protect terrestrial producers of metals."[17] What such economists would prefer to see, of course, are market forces guiding the nodule industry and expanding as long as it was profitable to do so. This system would result in the most efficient allocation of resources, it is claimed, and consumers would obtain minerals at the least possible cost. Eckert again makes his case, saying, "United States economic interests in ocean mining closely parallel those of economic efficiency."[18] U.S. Treasury officials, generally, have been of the same persuasion.

No one at Caracas was prepared to argue that economic efficiency was not in the narrow economic interests of the United States. What was at issue, however, was whether economic efficiency was a standard that would be in the global interests. Consensus formed among the developing world that it was not. Part of the problem here pertains to the time frame one adopts. Free-enterprise advocates claim that in the long term, consumers everywhere benefit from economic efficiency. This belief may ultimately be true, but few national representatives from developing nations feel they can spend the time waiting patiently for economic benefits to trickle down. Were the globe made up of states of equal economic might and capacity, the theories of free-market economists might have some relevance to the issue at hand. Clearly, however, the world is less orderly and less amenable to the conceptual framework of the economist. Though this fact was recognized by nearly all at the UNCLOS Caracas meeting, it did not dissuade the U.S. delegation from basing its case upon free-market economics. The U.N. Secretariat study stated the case against the U.S. position precisely: "The model of resource allocation favored by Western economists proceeds from a number of assumptions which are not all met in reality. It is obvious that perfect competition, free resource movement, and general access to technology are not true of the nodule industry. Also it must be recalled that the competitive model of resource allocation presupposes that the existing income distribution is socially and politically acceptable."[19]

To move beyond ideology and look more deeply into which consumers might gain from extensive nodule mining, we can examine U.S. Geological Survey data on worldwide mineral import and export figures. The most current figures were for the years 1968-1971.[20] Specifically, what was pertinent were the interests of the developing nations concerning the metals found in the manganese nodule. The conclusions derived from the data are as follows:

1. Cobalt is not a widely used metal at present, with the relatively small amount that is exported traveling nearly exclusively to the industrial nations. Zaire and Zambia are the major exporters.

2. The international traffic in nickel is somewhat larger than in cobalt; yet the direct use of nickel by the developing nations is still rather small. Out of

eighty-four developing nations for which figures could be obtained, seventy-four of them imported fewer than 100 metric tons of nickel per year. The developing nation importing the largest volume of nickel, India, only imported 3,558 metric tons in 1969. This figure contrasts with import levels exceeding 100,000 metric tons for both Japan and the United States. Nickel is primarily an export of developed nations.

3. Manganese, like cobalt and nickel, is primarily used in the production of steel. And as with nickel, seventy-four developing nations out of a total of eighty-four imported fewer than 100 metric tons of manganese per year. Again, the import levels of industrial nations dwarfed those of developing nations. The export picture cannot be interpreted unambiguously because distinctions between ore and processed ore are not reflected in the figures. Nevertheless, twenty developing nations were obtaining foreign exchange from the export of some form of manganese. Gabon is the largest developing nation that exports manganese ore, exporting nearly 2 million metric tons in 1969.

4. Copper, of course, is used extensively for many purposes and constitutes a healthy proportion of nonfuel mineral trade. As such, it is directly imported by far more developing nations than the other major metals in manganese nodules that have already been reviewed. Of eighty-four developing nations, half imported between 100 and 55,000 metric tons of copper per year (thirty-five of these forty-two nations import fewer than 5,000 metric tons per year). The case can be made, therefore, that many developing nations have a significant consumer interest in copper and its uses. However, it must also be noted that a total of thirty-four developing nations received foreign exchange earnings from the export of a form of copper (the major ore producers being Chile, Peru, Zaire, and Zambia). Developing nations consequently have both producing and consuming interests in copper.

Whether seabed mining for copper will result in worldwide consumer benefits is subject to question. Since it is forecasted that copper from seabed mining will represent such a miniscule portion of total copper production (see Table 4), it is highly unlikely that seabed mining will have any effect whatsoever on the price level of copper. Consuming nations can expect that prices will be affected by traditional forces. (Unfortunately, where seabed mining may make the largest price impact—on cobalt—there is a very small market for the product in the developing nations.)

The argument can be made, and often is, that, as importers or consumers of finished products containing the four major seabed metals, developing nations will benefit indirectly from seabed production. It is impossible to place a reliable figure on the extent to which consumers will benefit in this regard. The argument should be approached with a good deal of skepticism, however, because of the generally low proportion of costs that raw materials constitute in the manufacture of finished products. A United Nation's study concludes,

Savings from lower prices of raw materials, which do not represent a large

> proportion of costs, tend to be absorbed by manufacturers in the form of higher profits or wages, and are seldom passed on in the form of lower prices to consumers. Moreover, the minerals in point are not important inputs into wage goods—major items of consumer expenditure—for which declines in prices would have important real income effects.[21]

The contention that the important consumer interests of the developing nations will be furthered by unfettered deep seabed mining simply does not hold up under analysis, at least in the time frame relevant to current national concerns. For most developing nations, deep seabed mining will have very little economic impact for some time. A sharp division of perceived interests between developing nations that produce and those that consume cannot be viewed as a likely future scenario. Since this argument is so weak, the call for production or price controls by key mineral-producing developing states is not likely to stir an argument among the Group of 77.

Recognition that deep seabed mining contains little of substantive economic interest to the vast majority of developing nations differs sharply from the perception held by the international community in the late 1960s. Ambassador Pardo in 1967 spoke of the vast riches of the seabed, which, if devoted to the common heritage of mankind, could make a substantial contribution to the economic development of poorer nations. This vision, no doubt overstated at the time, had a powerful influence in kindling the interests of developing nations in deep seabed mining. The initial interest, therefore, was almost purely economic.

However, when international consensus during the early 1970s settled on extended national jurisdiction over the seabed (with the nearly complete national appropriation of seabed hydrocarbons), the glittering vision of seabed wealth for the international community in general withered. We can legitimately debate whether, in fact, there was ever the magnitude of potential ocean wealth described by Ambassador Pardo available to the international community; but with agreement upon extended national jurisdictions, the debate becomes academic. His estimate of potential revenues reaching $5 billion annually, raised in his speech of November 1, 1967, was based upon the assumption that all hydrocarbon deposits on the continental margin seaward of 200-meter depths would be under international jurisdiction. We see that this is no longer a valid assumption.

There can be little doubt that much of the support expressed by developing nations for the concept of "the common heritage of mankind" has been based on the expectation of considerable pecuniary rewards derived from mining. The distribution of rewards based on national needs rather than national capabilities struck a responsive chord in the developing world. The question of what magnitude of proceeds will be available to the international community cannot now be answered with certainty, but expectations have decreased precipitously since the late 1960s. After examining various possible scenarios based upon varying rates of entry and alternative production of minerals, a U.N. economist stated,

> Regardless of the actual level of tax revenues from nodules by 1985, it will not transform the scenario of world resource distribution and financial avail-

ability for development promotion in developing countries. Even in the unlikely event that the high assumption proved correct and all revenues collected from nodules were to be channelled to the developing countries, less than $0.30 per capita would be netted in 1985.[22]

One would intuitively think that the creation of an international organization to manage and regulate activities on the deep ocean seabed would be facilitated immensely by the rather meager resource pot that has now been alloted to the international community. In other words, it would seem probable that nations would compromise on international machinery now that the distribution of vast ocean wealth was not at stake. Contrary to such conventional analysis, however, the decreasing prominence of the resource issue per se has not produced accommodation. Why it has not will be discussed in subsequent chapters.

In summary, the urgency for seabed mining, stressed by mining companies and representatives from industrial nations, particularly from the United States, is now nowhere near as evident from the perspective of developing nations. One observer has claimed that the U.S. position at UNCLOS actually backfired and was a force in building G-77 unity.[23] As we have seen, developing nations are neither major consumers of the metals in question nor major importers. Moreover, seabed mining could possibly have a severe negative impact upon the export earnings of a few developing nations. Arguments, then, that "in the long run" seabed mining is in the interest of all nations have little impact. Any development that is likely to lead to an even greater income gap between developing and developed nations is hardly likely to find support within the broad international community today.

Notes

1. *"The Science, Engineering, Economics, and Politics of Ocean Hard Mineral Development,"* 4th Annual Sea Grant Lecture and Symposium, MIT, Oct. 16, 1975.
2. *Ibid.*, pp. 23–24.
3. *Ibid.*, p. 29.
4. *Economic Implications of Sea-Bed Mineral Development in the International Area: Report of the Secretary-General* (A/Conf. 62/25), United Nations, May 22, 1974.
5. Congressional Research Service, *Ocean Manganese Nodules,* Prepared for the Senate Committee on Interior and Insular Affairs, Washington, D.C., Feb. 1976, p. 47.
6. *A Report of the Conference on Economic and Related Impact of Manganese Nodule Mining,* The Stanley Foundation, Washington, D.C., Jan. 30–31, 1976, p. 4.
7. James C. Orr, *The Economic Effects of Deep Ocean Mineral Mining and the Implications for United States Policy,* Johns Hopkins University, Ocean Policy Project, Occasional Paper Series, No. 4, Dec. 1974, p. 25.

8. *Wall Street Journal,* October 30, 1974, p. 40.
9. *Economic Implications of Sea-Bed Mining in the International Area: Report of the Secretary-General* (A/Conf. 62/37), United Nations, Feb. 18, 1975, p. 5.
10. *Reports Submitted by the United Nations Conference on Trade and Development* (A/Conf. 62/26), June 6, 1974.
11. *Ibid.,* p. 1.
12. Statement by G. D. Arsenis, on behalf of the Secretary-General of the United Nations Conference on Trade and Development (A/Conf. 62/32), mimeo, n.d.
13. *Ibid.*
14. *Working Paper on the Economic Effects of Deep Sea-Bed Exploitation* (A/Conf. 62/C.1/L.5), August 8, 1974.
15. *Ibid.,* p. 10.
16. For this line of reasoning, see Richard J. Sweeney, Robert D. Tollison, and Thomas D. Willett, "Market Failure, the Common-Pool Problem, and Ocean Resource Exploitation," *The Journal of Law and Economics,* 17, 1 (April 1974), pp. 179–92; also in the same volume, Kenneth W. Clarkson, "International Law, U.S. Seabeds Policy, and Ocean Resource Development," pp. 117–42; and Ross D. Eckert, "Exploitation of Deep Ocean Minerals: Regulatory Mechanisms and United States Policy," in *Mineral Resources of the Deep Seabed,* Hearings before the Senate Subcommittee on Minerals, Materials, and Fuels, 93rd Congress, 2nd Session, March 11, 1974, pp. 1217–87.
17. Eckert, "Exploitation of Deep Ocean Minerals," p. 1255.
18. *Ibid.,* p. 1257.
19. *Economic Implications,* 1974, p. 79.
20. The data were collected by the Ocean Mining Administration, U.S. Department of the Interior, and generously made available to the author.
21. *Economic Implications,* 1975, pp. 7–8.
22. Raul Branco, "The Tax Revenue Potential of Manganese Nodules," *Ocean Development and International Law Journal* (Summer 1973), pp. 207–8. Branco assumes in his high estimate sixty-three mining ventures of 1 million tons per year production capacity in 1985 and a 30 percent tax revenue potential.
23. Edward Miles, "The Structure and Effects of the Decision Process in the Seabed Committee and the Third United Nations Conference on Law of the Sea" *International Organization,* 31, 2 (Spring 1977), 196.

Chapter 4

The United States Mineral Problem

Previous chapters have dealt with issues bearing directly on the mining of manganese nodules to acquaint the reader with its fundamentals. However, devoting exclusive attention to the specific enterprise would ignore the very real impact that broader mineral developments have had on the politics of seabed mining. The lenses through which national leaders now view mineral exploitation in general have changed from those of the 1960s. The movement of minerals in trade, once perceived solely in commercial terms, has recently become a matter of "high" politics. This chapter begins to explain why the straightforward technical task of raising and processing manganese nodules has become entangled in the resource politics of the 1970s. This chapter also seeks to demonstrate how corporate interests often become converted into national ones. The transformation of seabed mining from a "business" interest to a "public" interest has made accommodation more difficult to achieve.

The sense of urgency and negotiating rigidity that U.S. representatives to UNCLOS bring to the deep seabed issue cannot be fully explained without understanding the larger mineral situation facing the United States in the decade of the 1970s. In the early years of manganese nodule technological development, there was little indication that manganese nodule mining and its product would be viewed as anything other than a normal mining operation. To be sure, the ocean environment added a twist to the enterprise that drew special attention; the challenge and the novelty of it all were indeed exciting and noteworthy. But apart from the unique technology and environment, the task was essentially the same as that already being carried out throughout the globe.

However, Pardo's message in 1967 and the international response to it insured that deep seabed mining would be treated differently. The reference to seabed resources as the common heritage of mankind focused on the singularity with which these deposits were to be endowed. Yet before 1973-74 there was little in the testimony of U.S. government representatives or industry officials to indicate such uniqueness.

However, they did emphasize that domestic mining of manganese nodules would have a positive effect on the balance of payments. The United States at present is a major importer of the minerals found in the manganese nodule. Specifically, it imports almost 100 percent of its manganese needs, over 90 percent of its cobalt, between 80 and 90 percent of its nickel, and from 10 to 20 percent of its copper. Though less copper as a percentage is imported into the United States than the other three minerals, in terms of gross tonnage imported, it exceeds them. It has been estimated that in 1972 the total import bill came to over $1 billion.[1] Considerable attention, therefore, has been devoted to how manganese nodule mining could reduce the current balance-of-payments deficit. One study estimates that only three full United States production units bringing back and processing 100 percent of the four minerals in question could reduce manganese imports by 12 percent, copper imports by 41 percent, nickel imports by 53 percent; and cobalt imports could be wiped out altogether.[2]

Since 1973–74, however, this issue has taken a back seat to others that have proved to be far more prominent. During this period the "U.S. mineral problem" came to the forefront, and subsequently a link to manganese nodule mining was forged in the minds of many policy-makers. In other words, such mining often came to be viewed as a means of addressing the general U.S. mining malaise. One witness before Congress in 1974, for example, predicted an era of global mineral scarcity and competitiveness and therefore urged speedy congressional action: "Early exploitation of the resources of the sea, notably the nodules that lie in abundance in deep waters off our coasts, is therefore of great importance not only to our future economic health but also to our national security."[3] When viewed in such a light, manganese nodule mining becomes more essential than previously envisioned.

The "U.S. mineral problem" consists primarily of three dimensions. First, the diminishing stock of high-grade mineral ore found within the United States has led mining companies to search elsewhere for richer ore deposits, which in turn has led to an uncomfortable degree of dependence upon foreign sources for vital mineral resources. A greater degree of dependence on imports, therefore, is at the center of the perceived problem. Many observers also associate this growing reliance upon foreign sources with the thesis illuminated by the publication of the book, *The Limits to Growth.*[4] The book had its initial impact in 1973 and warned of impending shortages of life-supporting materials, including hard minerals, should industrial and population growth continue at then current rates. Although the chapter on possible mineral depletion was both simplistic and misleading, the thesis of the book as a whole was an intuitively persuasive one and gained a number of prominent adherents. The gradual depletion of high-grade U.S. mineral deposits was sometimes used as confirmation of the thesis. The abundance of manganese nodules, therefore, took on a larger importance than it had in the past, since mining the vast seabed deposits could not only reduce import dependence but also assure a source of minerals not subject to depletion within any relevant time frame.

Second, the vulnerability associated with mineral dependence was illustrated by the 1973 Arab embargo of oil to the United States. Moreover, the Organization of Petroleum Exporting Countries (OPEC) was subsequently able to raise oil prices at will. The successful OPEC cartel raised the hopes of other mineral-producing nations and led to considerable discussion about whether oil (and OPEC) was to be the exception in mineral exchange or the rule.[5] Intense concern was demonstrated in this country over the potential for collusion from suppliers. Manganese nodules and the deep seabed came to be viewed by many as an area that should remain free from such influences.

Third, since the mid-1960s, the role of private mining corporations, many based in the United States, has been seriously challenged globally. Foreign governments have demonstrated an increasing willingness to participate more fully in mining activities, capture an increasing share of the economic rent arising from mining, and finally, if necessary, expropriate mining assets. With the diminishing role of American-based private companies abroad, the less secure and assured becomes the timely and reliable supply of minerals. There has been widespread support within U.S. policy circles, therefore, for the goals and interests of the private mining corporations.

It is worth examining these three dimensions of the mineral problem more carefully to understand the major impact it has had in shaping U.S. seabed policy.

Declining Mineral Grade of Deposits

The United States is still the largest mineral-producing nation and is richly endowed with most of the minerals essential to an industrial society. Nevertheless, since the early 1900s, the United States has become increasingly dependent upon foreign mineral imports. Beginning in the 1920s, the United States went from a net exporter of minerals to a net importer. By 1972, it was only exporting one mineral in primary form—molybdenum—and the balance of trade for minerals and processed materials of mineral origin reached a deficit of $3.3 billion.

The depletion of easily mined domestic deposits is, of course, the major factor in mineral dependency. U.S. copper deposits, which have been intensively mined, now contain less than 1 percent copper, whereas foreign deposits often contain 3 percent or better. At the beginning of this century, the areas mined for copper in this country often contained yields of 4 percent or better; by 1970, the average yield of U.S. ores was only .65percent.[6] Another example is iron ore, whose deposits in the United States now average 19 percent, as compared to 39 percent worldwide.[7]

For this reason, and for other reasons of cost not directly associated with the grade of resources, the United States increasingly has looked abroad to meet the demands of industrial growth. Table 7 shows the percentage of U.S. mineral requirements imported during 1972 and from which countries.

In the absence of a direct policy change, U.S. dependence is expected to increase substantially throughout the rest of the twentieth century. Estimates of the import bill covering only nonfuel minerals by the year 1985 have been set at

TABLE 7. Percentage of U.S. Mineral Requirements Imported During 1972

MINERAL	PERCENTAGE IMPORTED	MAJOR FOREIGN SOURCES (THIRD-WORLD COUNTRIES IN ITALICS)
Platinum-group metals	100	U.K., U.S.S.R., South Africa, Canada, Japan, Norway
Mica (sheet)	100	*India, Brazil, Malagasy*
Chromium	100	U.S.S.R., South Africa, *Turkey*
Strontium	100	*Mexico,* Spain
Cobalt	98	*Zaire,* Belgium, Luxembourg, Finland, Canada, Norway
Tantalum	97	*Nigeria,* Canada, *Zaire*
Aluminum (ores and metal)	96	*Jamaica, Surinam,* Canada, Australia
Manganese	95	*Brazil, Gabon,* South Africa, *Zaire*
Fluorine	87	*Mexico,* Spain, Italy, South Africa
Titanium (rutile)	86	Australia
Asbestos	85	Canada, South Africa
Tin	77	*Malaysia, Thailand, Bolivia*
Bismuth	75	*Mexico,* Japan, *Peru,* U.K., *Korea*
Nickel	74	Canada, Norway
Columbium	67	*Brazil, Nigeria, Malagasy, Thailand*
Antimony	65	South Africa, *Mexico,* U.K., *Bolivia*
Gold	61	Canada, Switzerland, U.S.S.R.
Potassium	60	Canada
Mercury	58	Canada, *Mexico*
Zinc	52	Canada, *Mexico, Peru*
Silver	44	Canada, *Peru, Mexico, Honduras,* Australia
Barium	43	*Peru,* Ireland, *Mexico,* Greece
Gypsum	39	Canada, *Mexico, Jamaica*
Selenium	37	Canada, Japan, *Mexico,* U.K.
Tellurium	36	*Peru,* Canada
Vanadium	32	South Africa, *Chile,* U.S.S.R.
Petroleum (includes liquid natural gas)	29	*Central and South America,* Canada, *Middle East*
Iron	28	Canada, *Venezuela,* Japan, Common Market (E.E.C.)
Lead	26	Canada, Australia, *Peru, Mexico*
Cadmium	25	*Mexico,* Australia, Belgium, Luxembourg, Canada, *Peru*
Copper	18	Canada, *Peru, Chile*
Titanium (ilmenite)	18	Canada, Australia
Rare earths	14	Australia, *Malaysia, India*
Pumice	12	Greece, Italy
Salt	7	Canada, *Mexico, Bahamas*
Cement	5	Canada, *Bahamas,* Norway
Magnesium (nonmetallic)	8	Greece, Ireland
Natural gas	9	Canada
Rhenium	4	West Germany, France
Stone	2	Canada, *Mexico,* Italy, Portugal

SOURCE: National Commission on Materials Policy, *Material Needs and the Environment Today and Tomorrow,* Report for the U.S. Department of the Interior (Washington, D.C.: U.S. Government Printing Office, 1973), pp. 2–25.

approximately $20 billion, and by the turn of the century $44 billion.[8] Table 8 shows how U.S. dependence on key mineral imports has changed over time and is expected to evolve in the future.

The decline in self-sufficiency is clear and on its face alarming, but many observers never go beyond the obvious message to produce a careful analysis of the extent and nature of this dependency. The fact that mining increasingly has moved abroad does not, per se, indicate a lack of available domestic resources. The decision to mine abroad has essentially been made by corporate officials who seek to minimize the economic costs of production. Because costs are tied closely to the richness of the ore body, mining corporations have increasingly moved to areas where only limited industrialization has proceeded and where many rich ore bodies are intact. Therefore, the effective or real dependence of the United States upon foreign imports is decreased by the knowledge that extensive deposits (albeit low grade ones) still exist and could be mined should one want to pay the costs involved. A study by the Library of Congress, for example, has estimated that the United States could, given adequate incentives, become self-sufficient in fifty of sixty-three separate materials now used in the industrial process.[9]

It is instructive to look briefly at the mineral dependence of other industrial nations in comparison to the United States. Many of them, such as Japan, the Federal Republic of Germany (FRG), and France, have virtually no alternative to large-scale dependency because of their limited resource base. Both Japan and the FRG, for example, import 90 percent or more of their petroleum, copper, nickel, iron ore, and bauxite. Because these nations have never possessed the resources to support an industrial society or have exhausted most of the deposits

TABLE 8. U.S. Dependence on Imports of Principal Industrial Raw Materials with Projections to Year 2000 (Percent imported)

RAW MATERIAL	1950	1970	1985	2000
Aluminum	64	85	96	98
Chromium	n.a.	100	100	100
Copper	31	0	34	56
Iron	8	30	55	67
Lead	39	31	62	67
Manganese	88	95	100	100
Nickel	94	90	88	89
Phosphorus	8	0	0	2
Potassium	14	42	47	61
Sulfur	2	0	28	52
Tin	77	n.a.	100	100
Tungsten	n.a.	50	87	97
Zinc	38	59	72	84

SOURCE: Lester R. Brown, *World Without Borders* (New York: Random House, 1972), p. 194; reprinted by permission of the publisher. Data are derived from U.S. Department of the Interior publications.

they once had, they have been forced to accept dependency as a condition of their economic growth. The Soviet Union, however, which is only minimally dependent upon mineral resources outside its borders, has a storehouse of high-grade deposits waiting to be exploited.[10]

Thus the United States represents an intermediate case between its resource-poor industrial allies and the resource-rich Soviet Union. Should security considerations loom paramount in the mining effort, the United States could move back toward a position of self-sufficiency.

It is not only the economics of mining low-grade deposits that is preventing large-scale mining in the United States but also the social and environmental costs. As one mining official has stated,

> An attempt to achieve self-sufficiency would require a heavy national commitment. It would be necessary to maintain high prices, to provide federal subsidies, to restrict imports, to relax environmental protection measures, and to raise staggering sums of money to finance new mineral production. In the absence of a crisis, it is doubtful that much support will be generated for such an extreme program.[11]

At no other time in history have U.S. citizens appeared quite so unprepared to bear such burdens. If the current energy situation can legitimately be termed a "crisis," we can see that even in such a predicament general dedication to achieving self-sufficiency has been lukewarm at best. We can assume that pleas for support of even broader self-sufficiency would occasion even less enthusiasm. We would also be working, as a nation, at cross-purposes. As lower and lower grades of ore are mined, increasing energy consumption per unit of output is required.[12] One cost of a program devoted to national self-sufficiency in nonfuel minerals, therefore, would be higher energy consumption—clearly a consequence at variance with national goals.

Mining in this country has come to be viewed by the public as a less than desirable activity. The extraction, processing, and refining stages of mining are all serious polluters of air and water. Consequently, state and federal legislation for environmental standards has been applied to all mining activities. The additional capital investment costs of meeting environmental standards have been substantial, and they have been a major factor in the shift of many mining enterprises to foreign mines—where environmental standards are either less restrictive or nonexistent. In addition, there is increasing pressure for massive surface reclamation following extraction, which will also involve substantial added-on costs.

There is also much sentiment expressed for the preservation and conservation of remaining wilderness areas. Legislation, in response to such sentiment, is closing off public lands to mining, and thereby restricting exploration for new and rich mineral ores.

One hears a great deal of the price squeeze that foreign governments have enacted upon multinational corporations (MNCs), but surprisingly little about similar actions by state and local governments in this country. Western states in particular have imposed heavy taxes on mineral activities,[13] and several governors have now made it clear that their states will no longer bear a disproportionate share of the costs of energy and mineral extraction.

In summary, therefore, a movement to reverse the trend and level of U.S. mineral dependence would run directly counter to existing pressures, which tend to encourage greater use of imports. The tenacity of these existing pressures should not be underestimated. In the face of a most precarious energy situation, environmental forces in this country have yielded little to those who would lessen environmental standards. Without a most extreme situation, therefore, a program aimed at mineral self-sufficiency would create a divisive and antagonistic atmosphere throughout the United States. The hazards and uncertainties of dependence may very well be more acceptable than the price that must be paid for autarchy.

Manganese nodule mining is viewed as a means of bypassing the difficult choices inherent in either a policy of import dependence or mineral autarchy. Manganese nodules can bring self-sufficiency over time for four select minerals; but four minerals cannot, in themselves, reverse the long-term trend toward greater U.S. mineral dependence nor significantly reduce balance-of-payments deficits. However, manganese nodules have symbolic value in that their appeal is to an age past, an age when minerals were abundant, mining operations were "invisible," and there was no need to be dependent upon the whims of foreign governments.

Supply Manipulation

The hike in crude oil prices by the OPEC cartel in 1974 had a significant effect upon other mineral-producing nations, as it illustrated the very substantial benefits that could accrue to suppliers through collusion. The twelve-nation oil cartel, by boosting the price of crude oil, had raised its collective revenues from $25 billion in 1973 to $80 billion in 1974. Seeing such a change, the producers of other commodities began to evaluate their chances of forming an effective cartel, or at least capturing a greater share of the economic rent from mining. There was, consequently, considerable public debate in the United States over whether the oil cartel was an exception or would become representative of the norm. Whether the traditional "boom and bust" cycle so typical of the mineral market was to be replaced by a carefully structured and controlled price cycle was part of the debate. One economic analyst claimed there had been a basic transformation from a buyer's to a seller's market: "Shortages of supply have replaced shortages of demand as the dominant force in world economics for the first time in almost 50 years, and the power position of suppliers and consumers has thus changed dramatically."[14]

For many reasons, prices for most minerals were high in 1974. Also at this time various mineral producers made significant efforts to increase their export earnings: Seven prominent bauxite exporters formed the International Bauxite Association (IBA), and one member, Jamaica, demanded and received a sixfold increase in its export earnings. The major copper producers of the Intergovernmental Council of Copper-Exporting Countries (CIPEC) agreed upon an export quota system which would involve a reduction in copper exports. Tin producers were demanding a sharp increase in the internationally guaranteed floor price maintained for this commodity. Iron ore producers held a conference to discuss means of coordinating price strategies. Phosphate producers tripled the price of phosphate and were expecting to raise the price even further. Finally, the special session on raw materials held by the U.N. General Assembly in 1974 called upon developing nations to form commodity associations in order to exercise full "sovereignty" over their natural resources.

The possibility that these efforts and others would be successful raised considerable fear in the industrial nations. The prospect of a competitive struggle over "scarce" resources was fraught with danger, as revealed by Helmut Schmidt, chancellor of the FRG:

> The struggle over oil prices may be followed by a similar struggle over the prices of other important raw materials. And since what is at stake is not just pawns on a chessboard, but the peaceful evolution of the world economy and the prosperity of the nations of this world, we need a politically sound philosophy if we are to win this dangerous fight.[15]

Reaction to the rhetoric of "resource politics" in this country was often heated and intensely nationalistic. The OPEC cartel in itself was painful enough, and the prospect of other cartels was bitterly opposed. In reality, though much discussion continues on the international level, the budding of prototype OPECs has failed to materialize since 1974, and mineral prices have fallen considerably. The "worst-case" scenarios, which various U.S. observers had predicted, simply have not occurred.

Even in 1974 there were knowledgable observers who emphasized the difficulties involved in maintaining a united cartel over a projected time period.[16] There are obvious problems in uniting a large group of culturally and politically diverse producers, each of which is in a unique market position and is looking for a way to increase export earnings at the expense of others. The potential for mineral substitution is also a deterrent to large price hikes. For these and other reasons the proliferation of cartels has never come to pass.

It should also be pointed out that developing nations—which because of their relative poverty might be more politically and economically motivated to raise mineral prices—generally do not possess controlling shares of the various mineral markets. One of the enduring common wisdoms is that the world is neatly divided between industrial countries that import mineral commodities and the nonindustrial developing world that exports them. In reality, in 1970 developing nations accounted for only 38 percent of fuel and nonfuel mineral exports in terms of value.[17] With the continuing rise in the price of petroleum, this percentage has now increased, but the portion attributed to nonfuel mineral commodities has remained relatively stable. The developing world possesses no more than 45 percent of the world's known reserves of nonfuel minerals.[18] The data in Table 9 show that developing nations capture a major share of the world export market for only a relatively few important minerals, including crude petroleum, bauxite, tin, and manganese ore.

Petroleum occupies a special place in world trade because it is a unique and widely used commodity. Energy is an integral part of all industrial processes and transportation. There are no readily available substitutes for petroleum that can be brought on line easily and without unfortunate economic side effects. Aside from petroleum, only copper and iron could honestly be referred to as "building blocks" of industrial civilization. For most other minerals adequate replacements could be developed, given requisite lead time and willingness to undergo some economic hardships.

The 1974 increases in petroleum prices brought its share (by value) in world trade from 7.7 percent to nearly 25 percent.[19] At present, the value of petroleum exports in international trade nearly equals that of all basic foodstuffs and other extractive minerals combined. The OPEC subgroup of developing nations is the primary beneficiary of these price increases. In 1970, OPEC countries captured 89 percent of the world export market. In terms of total exports from developing countries—both raw materials and manufactured products—a recent U.N. study estimates that in 1974 petroleum accounted for no less than 62 percent of their receipts.[20]

The critical point emerging from these data is that petroleum revenues accrue to only selected developing countries, and the rising price of this commodity places an added burden on the vast majority of other developing countries. Providence has not favored all nations equally in distribution of mineral resources. Most developing nations have either extremely small or nonexistent

TABLE 9. Developing Countries' Share of Major World Mineral Exports

MINERAL	PERCENTAGE	MINERAL	PERCENTAGE
Crude Petroleum	89	Nickel Ore	24
Bauxite	88	Chrome Ore	22
Tin	77	Zinc Ore	14
Tin Ore	64	Zinc	12
Manganese Ore	51	Lead Ore	12
Petroleum Products	45	Lead	11
Copper	44	Nickel	7
Copper Ore	42	Aluminum	5
Iron Ore	42	Coal	3

SOURCE: "Study of the Problems of Raw Materials and Development: The Significance of Basic Commodities in World Trade in 1970" (A/9544/Add. 1), United Nations, Apr. 4, 1974.

mineral sectors and are net mineral importers. It is estimated that sixty-one out of 103 developing nations have only a small mineral sector or none at all.[21] In the third world roughly 90 percent of petroleum and other mineral exports are accounted for by countries containing less than one-fourth of the population.[22] Therefore, one of the major reasons that developing nations capture only a small share of most mineral export markets is that there are relatively few of them involved in mineral production on a large scale. For three-fourths of the population of the developing world, increasing mineral prices constitute an additional liability rather than an asset.

The differing scales of mineral exports and the importance of them to the economies of developing nations are illustrated in Table 10. Once again, the prominence of petroleum is particularly striking. As of December 1973, eighteen developing nations were obtaining more than 50 percent of their export earnings through the sale of a single mineral; but only five of these nations obtained the bulk of their export earnings from minerals other than petroleum. Outside of petroleum, and copper in Zaire, Zambia, and Chile, mineral commodities do not play the terribly significant role in export earnings that is commonly ascribed to them. There are several developing countries earning a significant portion of revenues from the export of one nonfuel mineral, but most of these nations are only earning 10 to 30 percent of their revenue from this source.

These data indicate that in the long term the fortunes of most third-world countries do not necessarily rise with increasing mineral prices. An initial flush of political success with petroleum prices has recently bolstered third-world solidarity, but a more sober assessment will undoubtedly reveal little economic "trickle down" in the international system.

Thus, analyzing the future prospects and patterns of interdependence in the third world is more difficult and complex than would initially seem. A clear hierarchy of nations exists in terms of potential mineral trade, OPEC countries standing quite apart at the top. Although OPEC members have shown some willingness to make loans and grants to other developing nations, a global cost-benefit analysis would surely reveal that their actions have cost most nations much more than they have gained.[23] The newly rich OPEC countries have acquired political power through the transformation of old mineral relationships, which has not been shared with other third-world nations. It remains to be seen if the new political and economic power is used to further collective third-world interests.

Beneath the layer of rich petroleum producers is a smaller number of nonfuel mineral producers who could conceivably profit from any general escalation of prices. Zambia, Chile, Zaire, Jamaica, Mauritania, and several lesser exporters of minerals fall into this second tier of the third world. But although the future could hardly be described as bleak for these countries, they have little prospect of ever attaining considerable political and economic power in their own right. In fact, it is questionable whether increases in the price of their nonfuel exports could ever be substantial enough to make up for what has been lost through increased bills for fuel imports.

TABLE 10. Percentage of Export Earnings from Single Mineral Commodities (1973)

	90%	70–90%	50–70%	30–50%	10–30%
Developing Nations Exporting Crude Petroleum and Petroleum Products	Venezuela Saudi Arabia Kuwait Libya Iran Iraq Abu Dhabi Qatar	Nigeria Yemen (Peoples Rep.) Trinidad-Tobago	Algeria Indonesia	Gabon Tunesia Ecuador	Panama Singapore Bolivia
Developing Nations Exporting Mineral Commodities Other Than Petroleum	Zambia (copper)	Chile (copper)	Zaire (copper) Mauritania (iron ore) Jamaica (alumina-bauxite)	Bolivia (tin) Guyana (bauxite) Niger (uranium) Togo (phosphates)	Morocco (phosphates) Peru (copper) Gabon (manganese) Haiti (bauxite) Jordan (phosphates) Malaysia (tin) Senegal (phosphates) Philippines (copper) Tunisia (phosphates) Sierra Leone (iron ore) Rwanda (tin)
Total Number of Developing Countries	9	4	5	7	14

SOURCE: Compiled from the IMF International Financial Statistics, December 1974.

In summary, the idea that many U.S. observers had in 1974 that all developing nations were both able and willing to forge commodity cartels was based upon a false impression of third-world mineral strength and cohesion. Cartels work against the interests of the vast majority of developing nations, especially those with only a small mineral sector which they hope to expand in the future. Copper is a particularly good case in point. Peru, Chile, Zaire, and Zambia are the major copper exporters among the developing countries. Several other third-world nations now produce only small amounts of copper but would like to expand production in the future—countries such as Mexico, the Philippines, Iran, Brazil, and Panama. Any scheme that regulates or restricts prices or stock flow would tend to cement the dominant position of the major copper exporters at the expense of those with few exports but high potential.

For example, the United States is currently not greatly dependent on mineral sources found in the developing world. As a whole, approximately two-thirds of all U.S. mineral imports come from developed countries, with Canada and Australia the primary sources. Of the imports from third-world countries, there is in general a diversification of suppliers, thereby eliminating excessive reliance upon individual nations.[24] Much of the fear over cartels in this country during the years 1974-75, therefore, was based upon an extrapolation of the oil cartel to other mineral commodity groups. As we have seen, however, oil represents a unique commodity and set of producers, unlikely to be easily replicated.

Fear of a Manganese Nodule Cartel

Manganese nodules became embroiled within the larger debate on resources after the successful oil embargo and OPEC price hike. Whereas in pre-1974 days attention centered on the balance of payments benefits associated with deep seabed mining, post-1974 statements inevitably cited "assured access to manganese nodules" as a nonnegotiable demand. Spokesmen for the leading nodule mining entities were only too happy to find an argument so clearly in the "public interest" with which to agree. Not untypical of the sentiment expressed at congressional hearings were the words of House Merchant Marine and Fisheries Chairman John M. Murphy:

> American industry cannot—and should not—commit hundreds of millions of dollars to the recovery of those minerals in the face of an international cartel which hopes to control every facet of ocean mining at our expense.[25]

A strong and persistent theme of opposition to the creation of a strong seabed international organization, which would have significant representation by developing nations, has been the fear that it would, in collusion with land-based producers, either deny minerals altogether to the United States or hike prices. Such fears, as we have seen, are based more upon emotion than hard analysis. A careful analysis shows that there is less reason for concern than some would think–which can be seen in a brief review of the four major minerals found in manganese nodules and the U.S. import position.

1. *Nickel:* The United States imports approximately 85 percent of its consumption, which is a significant amount. The suppliers, however, are primarily other industrial developed nations–mainly Canada and Norway–which are unlikely to impose new export conditions or restrictions.

2. *Manganese:* Over 90 percent of U.S. consumption is imported, much of it from developing nations like Brazil and Gabon. Thus far there has been no hint of supplier collusion, perhaps because manganese is one of the most abundant minerals on earth and South Africa and Australia contain the world's largest reserves. As mentioned previously, the manganese in nodules is pure manganese which cannot be directly substituted for ferromanganese. France and South Africa supply over 75 percent of our ferromanganese imports.[26] Just as for nickel, the supply links to the United States seem reasonably reliable and secure.

3. *Copper:* Major copper producers from developing nations have shown the greatest interest in price manipulation of all the nodule-mineral suppliers. Peru, Chile, Zambia, and Zaire have been at the forefront of CIPEC's (Conseil Intergouvernemental du Pays Exportateurs de Cuivre) formation, but thus far have been rather unsuccessful in affecting price. Some of the reasons for this failure are the large number of competing copper suppliers which cannot be easily coopted into CIPEC and the availability of substitutes for copper. As for the United States, its imports of copper are relatively minimal, averaging 15 percent of consumption per year. As the largest mineral producer in the world, with large reserves yet remaining, the United States could become self-sufficient in copper if there were the perceived need. Thus where some suppliers have been most anxious to emulate the OPEC cartel, the United States has little to fear because of its domestic supplies and its ability to find substitutes for copper in some applications.

4. *Cobalt:* The United States has in the past imported as much as 90 percent of its consumption in a year. Zaire has been the major trading partner, supplying approximately 75 percent of U.S. imports. Should Zaire, for whatever reason, cut off U.S. supplies, there might be short-term inconveniences but little more than that. Nickel can be substituted for cobalt, and in addition, the United States has close to two years of cobalt stockpiled.[27] Finally, most cobalt is mined as a by-product of copper, nickel, and iron mining, and hence can be found in abundance throughout the world. The small volume of cobalt used lessens the possibility that action by suppliers will produce a significant effect.

One study of possible collusion by suppliers has come to the following conclusion, in which the author concurs: "The existence of diverse producers, the elasticities of demand for and supply of the four minerals and the implausibility of export discipline and policy cohesion among producers argue strongly against the possibility of sustained influence over the international market in any of the major minerals found in manganese nodules."[28]

The preponderance of evidence suggesting the unlikelihood of collusion, however, has not defused the issue as one might expect. Instead, general import dependence remains a rallying point for those more concerned with emotion than fact. Such dedication to "minerals rhetoric" has made accommodation more difficult to achieve at UNCLOS.

The Changing Role of Multinational Corporations

Since the establishment of industrial societies, mining has been carried out by large, often vertically integrated organizations, which because of their scale and level of expertise have played a predominant role in establishing mining rules and trade. Because of their unique combination of technological expertise and management skill, they often became powers in their own right. The fact that many of these organizations were based in the United States provided a sense of security over the continued supply of minerals, even when exploitation was taking place outside of national borders.

An indication that such security could no longer be taken for granted was provided by the Arab embargo of petroleum in 1973. Multinational corporations (MNCs) essentially carried out the orders of Arab officials to the detriment of American and Western European consumers. The large oil MNCs quite correctly understood that their continued existence in Arab nations depended upon closely cooperating with Arab representatives rather than with those from their home bases.

However, the Arab embargo was not the first indication that the standard MNC role in foreign countries was changing, and one can find evidence of it long before 1973. It is worth looking briefly at this evolution.

American, British, Belgian, and French MNCs have played a major role in the production, processing, and marketing of minerals found in developing countries. Long before manufacturing interests in the industrial nations had begun selling their wares across the globe, mineral-extracting MNCs were actively involved in the developing nations. These companies have provided four essential functions for developing nations: (1) They have supplied the large amounts of risk capital that are required for an enterprise of this nature; (2) they have supplied the technology and know-how required in both exploration and exploitation of mineral stocks; (3) they have provided high-level management, which is required for efficient operation; and (4) they have marketed the final product.[29]

In return for their capital, technology, and services, the MNCs have normally exacted "concessions" from developing nations, meaning that contracts are drawn up granting exclusive rights to specific MNCs to explore and exploit designated areas over a specified period of time. By securing concessions the MNCs were, in effect, locking up sources of mineral supply to ensure continued downstream production.[30] Under this system the role of the state in both the producing and consuming nations has been limited. The success of MNCs in obtaining highly favorable mineral concessions was no doubt abetted in the past by the general ignorance of foreign nationals in regard to the legal and economic particulars of mining.[31] With the "learning curve" of foreign nationals considerably higher today (assisted, no doubt, by their widespread education in this country), the opportunity for MNCs to obtain such favorable mining rights is considerably diminished.

Dissatisfaction with the concession system has been voiced both frequently and vehemently by representatives of developing nations. This dissatisfaction initially leads to a renegotiation of contract terms, and few concession agreements drawn up with developing nations since World War II have remained unaltered.[32] The most common demand from the host country in such renegotiations has been for a continually larger share of the economic rent derived from mining, and in most cases they have been successful. During the 1920s petroleum companies paid to host governments royalties of approximately 15 percent of the commercial value of the oil produced and no income tax. By late 1960 royalties and taxes had risen to a level between 50 to 75 percent of the net earnings.[33] A similar trend is found in nonfuel mineral industries, where taxes and royalties, previously negligible, now frequently exceed 50 percent of company revenues.[34]

During the past decade, however, dissatisfaction has led to initiatives more fundamental than simply a larger slice of the royalty pie. There has been a persistent and widespread movement to gain control of production processes, thus pressuring multinationals out of the market. This action is a part of the desired "new international economic order," whose essential principle is that every state should gain full and permanent sovereignty over its natural resources and all economic activities.[35] The growing nationalization of production in developing nations is evident with regard to several minerals. Until the mid-1960s, copper was under the almost exclusive control of large MNCs. Among the major copper exporters, Zaire was the first, in 1967, to totally nationalize production. Peru and Chile followed in the 1970s, and Zambia now has nationalized a large percentage of its domestic copper resources. Petroleum offers another example. "In less than half a decade (from 1970 to mid-1974), the traditional phenomenon of foreign majority-owned or controlled petroleum facilities has become the exception rather than the rule in the OPEC countries."[36] Bauxite-producing nations have begun taking similar steps.

The assertion of control by developing nations over mineral production is based on their judgment that the bundle of services and functions that the MNC has traditionally provided through concessions can now be obtained either from other sources or from the MNCs themselves, without concessions. New sources of capital are being pursued. Industrial nations have expanded the provisions of export credits on a far larger scale than previously, and several mineral-producing nations are now going to the private capital market to obtain necessary financing. In addition to the traditional sources of capital, oil-exporting nations now have the capacity to provide capital funding; there is evidence that several developing nations are expanding their mineral sector with OPEC capital.[37] There are also increasing pressures being placed upon such international institutions as the World Bank, the United Nations Development Fund, and regional development banks to increase considerably their financing of the development of natural resources.

In terms of technical know-how and management, many developing nations now feel that a sufficient indigenous base of expertise, or human capital, has been built up over the past two decades. Other developing nations now feel themselves to be in a position of strength and are approaching MNCs to perform mining service contracts (free from significant MNC equity). The MNC response to such overtures has thus far been guarded, but not necessarily adverse. The Cerro Verde copper project in Peru, controlled by Peru's State Mining Enterprise, has enlisted a consortium of MNCs to provide both engineering and management services. The success of this project is being closely observed by other nations.[38] Venezuela, which recently nationalized oil production, is currently debating whether to invite MNCs back under service contracts.

Finally, even the marketing function of MNCs is beginning to erode. The persistent movement among developing nations to gain control over the production of mineral resources greatly increases their discretion in regard to rates of mineral supply, the price of the commodity, and the ultimate recipients. On the other hand, resource importers have been only too happy to circumvent the multinationals and make state-to-state deals—the implications of which have been grasped by those industrial nations currently most dependent upon mineral imports. The Japanese, French, and West Germans have been very active in securing guarantees of mineral supplies from developing nations, sometimes in exchange for technical and capital assistance. These deals are often bilateral, involving the exchange of technology for minerals, or multilateral, such as the 1974 agreement between EEC members and former colonies. The Japanese, with their pressing need for guaranteed access to foreign minerals, have been especially aggressive in trading technological know-how for long-term deliveries of various vital minerals.

However, the United States has been hesitant to initiate bilateral agreements for long-term supplies of minerals in exchange for technical assistance. This reluctance results from the fact that there exists no state mechanism to engage in such trade, and there is strong conviction that mineral exchange should remain the province of MNCs. U.S. officials have repeatedly sung the praises of a free market for mineral commodities even though such a free market has never in fact existed. The fact is that the United States is only belatedly discovering what other resource-deficient nations have known all along: that the interests of private corporations do not necessarily coincide with those of the state. Nor can one nation successfully play at free enterprise while the rest of the world is increasingly relying upon bilateral and multilateral agreements for mineral security. The fact is that mineral trade is becoming politicized, and there is increasing pressure for both producers and consumers to intervene in world markets.

In addition to government intervention in both exporter and importer countries, MNCs are increasingly caught in a squeeze that is considerably decreasing their freedom of action. On the one hand they are watching their concessions disappear in one developing country after another as strident nationalist groups demand confiscation of foreign assets. On the other hand, they are increasingly feeling pressures from industrial countries to act in the national interest. Moreover, the position of multinationals as processors and distributors of mineral commodities is gradually being undercut by government-to-government arrangements for the export of processing technology.

The willingness of the Japanese and Western European nations to provide capital without corporate investment in exploration and production, combined with third-world aversion to direct foreign control, produces a very new mining atmosphere, institutionally. This atmosphere portends, not the elimination of the MNC in resource development, but a radical transformation of its role. It is likely that MNCs will increasingly be called upon to fulfill specific service contracts for exploration and exploitation, with an elimination of most, if not all, equity investment opportunities.

UNCLOS and MNCs

Negotiations at UNCLOS have demonstrated that even when one goes to the deepest floor of the oceans, one cannot escape the forces and factors that shape terrestrial practices. Control over MNC activity in exploitation of ocean resources has in fact been the major, pervasive theme of UNCLOS. Except for the socialist nations, it is largely incorrect to claim that nations as such have exploitation capabilities. Instead, MNCs generally develop, own, and use the technology necessary for exploitation—in fishing, in oil and gas exploitation, and in the budding manganese nodule industry.

The primary means devised to control MNC activity in the ocean became obvious early in the negotiations, namely, to extend national jurisdiction over resources far out from the coastline. The proposal for an exclusive economic zone (EEZ) extending national jurisdiction over ocean resources beyond 12-mile territorial seas and out to 200 nautical miles from shore originated with several key Latin American delegates, and in the early 1970s gained support from numerous African and Asian delegations.[39] Most industrial nations have now, rather grudgingly in some cases, acquiesced in the EEZ concept; and it appears likely that national jurisdiction over resources to 200 nautical miles from coastlines will become customary law, if not also treaty law. This extension of national jurisdiction is extremely significant, since approximately 87 percent of the potentially recoverable oil and gas is found shoreward of 200-mile extensions as well as 90 percent of the fish stocks. It also appears that any future oceanographic research taking place within EEZs and dealing directly with resources will require the prior consent of the appropriate coastal nation. In short, the conditions under which the vast bulk of ocean resources will be developed will be dictated by states and not left to the autonomous discretion of MNCs.

Although industrial nations have accepted the EEZ proposal incorporating state control over MNCs, they have steadfastly refused to go along with the wishes of developing nations to place strong international control over MNC activities on the seabed beyond national jurisdiction. This subject will be discussed in subsequent chapters; it is enough to say here that many industrial nations, particularly the United States, have not accepted the possibility that anything deviating far from the traditional concessional approach to mining is appropriate for deep seabed mining. Hence, whereas the terms of terrestrial mining have changed radically over the past decade, and whereas state control over most ocean resources is now accepted, many countries still hold that the deep seabed area should remain the province of MNCs. In fact, it appears that the more that land-based mining rules and operations change, the more insistent many in the United States become that the seabed *not* follow the same trends. This sentiment was colorfully expressed in the statement made by one proponent of unilateral U.S. seabed mining: "The risk of appropriation or failure of reliable production through political unrest does not hang, as a sword of Damocles, over the marine source of supply."[40] The deep seabed, therefore, is sometimes viewed as a last bastion of the MNC preserve, not to be conceded at any cost.

Conclusion

The U.S. minerals problem has greatly retarded significant progress in UNCLOS negotiations on the seabed. The precariousness of our industrial society, based upon the dependence on imported materials, is a fact to which everyone can relate. The dramatic convergence of the three strands that constitute the mineral problem left an indelible impression on the public conscience, specifically on the minds of many elected officials; but the thesis of this chapter has been that the U.S. mineral problem is far more apparent than real. The limits-to-growth thesis as it pertains to nonfuel minerals has been cogently and persuasively rebutted.[41] There is no justified fear that we will run out of nonfuel minerals. Specific U.S. mineral dependence is still a choice rather than a necessity, and general dependence is neither large nor based on key exporters from developing nations. Commodity manipulation has rather fizzled out since the scare of 1974, revealing the difficulties involved in putting producer cartels together. Nor is it likely that the four key minerals in the manganese nodule could become subject to a general supplier cartel. Finally, although the rules of the game have changed significantly for MNCs over the past decade, there is no indication that either they or the United States will be unable to live with the new rules. MNCs are now active throughout the globe working for governments of every ideological stripe and functioning effectively in various capacities. In short, the so-called mineral problem has unnecessarily retarded negotiating progress at UNCLOS and threatened the collapse of the entire conference. More reflection by key officials on the actual dimensions of the problem is clearly warranted.

Notes

1. Robert M. White, "Sounding Our Ocean Future," Address to the National Oceanic and Atmospheric Administration Conference on the Oceans and National Economic Development, Seattle, Wash., July 17, 1973.
2. Congressional Research Service, *Ocean Manganese Nodules,* Prepared for the Senate Committee on Interior and Insular Affairs, Feb. 1976, p. 47.
3. Robert A. Kilmarx, *Mineral Resources of the Deep Seabed, Part 2,* Hearings before the Senate Subcommittee on Minerals, Materials, and Fuels, 93rd Congress, 2nd Session, March 1974, p. 897.
4. Donella H. Meadows *et al., The Limits to Growth* (New York: Universe Books, 1972).
5. See C. Fred Bergsten, "The Threat is Real," and Stephen D. Krasner, "Oil is the Exception" in *Foreign Policy,* 15, (Spring 1974), 84–89 and 68–83, respectively.
6. Statement by Stephen A. Wakefield, *National Materials Policy,* Hearings before the Senate Subcommittee on Minerals, Materials, and Fuels, 93rd Congress, 1st Session, November 1973, p. 95.
7. Alvin Kaufman, "Are We Running Out of Mineral Technology?" *Mining Engineering,* 24, 9 (Sept. 1972), 54.
8. *Mining and Minerals Policy 1973: Second Annual Report of the Secretary of the Interior under the Mining and Minerals Policy Act of 1970* (Washington, D.C.: Government Printing Office), 1973.
9. Congressional Research Service, *Report on Raw Material Resources,* reprinted in the *Congressional Record,* April 22, 1974, p. S6006.
10. Of the thirty-six minerals crucial to industrial processes, the Soviet Union lacks self-sufficiency in only ten. Nazli Choucri and James P. Bennett, "Population, Resources and Technology: Political Implications of the Environmental Crisis," *International Organization* (Spring 1972), p. 196.
11. Clayton J. Parr, "Non-fuel Minerals: Another Dimension to the Problem of Resource Availability," Paper presented to the 20th annual Rocky Mountain Mining Institute, San Francisco, Calif., July 11, 1974, p. 11.
12. David B. Brooks and P. W. Andrews, "Mineral Resources, Economic Growth, and World Population," *Science,* 185, 4145 (July 5, 1974), 18.
13. Parr, "Non-Fuel Minerals," p. 16.
14. Bergsten, "The Threat is Real," p. 85.

15. Helmut Schmidt, "The Struggle for the World Product," *Foreign Affairs* (April 1974), p. 438.
16. See Bension Varon and Kenji Takeuchi, "Developing Countries and Non-Fuel Minerals," *Foreign Affairs* (April 1974), pp. 497–510; and Raymond F. Mikesell, "More Third World Cartels Ahead?" *Challenge* (Nov./Dec. 1974), pp. 24–31.
17. U.S. Department of the Interior, *Minerals Yearbook,* 3 (1971), 6.
18. Varon and Takeuchi, "Developing Countries," p. 508.
19. "Study of the Problems of Raw Materials and Development: The Hypothetical Impact of Commodity Price Movements on World Trade" (A/9544/Add. 2), United Nations, April 12, 1974, p. 2.
20. Raymond F. Mikesell, ed., *Foreign Investment in the Petroleum and Mineral Industries* (Baltimore: Johns Hopkins University Press, 1971), p. 6.
21. Andrew J. Freyman, "Mineral Resources and Economic Growth," *Finance and Development* (Mar. 1974), p. 21.
22. *Ibid.*
23. "Study of the Problems of Raw Materials and Development," p. 3.
24. Leslie H. Gelb, "U.S. Study Hopeful on Raw Materials," *The New York Times,* Nov. 13, 1974, p. 48.
25. Congressman John M. Murphy, news release, March 16, 1976.
26. Richard C. Raymond, "Seabed Minerals and the U.S. Economy: A Second Look," *Marine Technology Society Journal,* 10, 5 (June 1976), 16.
27. *Ibid.,* p. 17.
28. *Ibid.*
29. Mikesell, ed., *Foreign Investment,* pp. 26–27.
30. Theodore H. Moran, "New Deal or Raw Deal in Raw Materials," *Foreign Policy* (Winter 1971–72), pp. 122–23.
31. Theodore H. Moran, "The Theory of International Exploitation in Large Natural Resource Investments," in Steven J. Rosen and James R. Kurth, eds., *Testing Theories of Economic Imperialism* (Lexington, Mass.: Lexington Books, 1974), p. 167.
32. *Ibid.*
33. Raymond F. Mikesell, "Healing the Breach Over Foreign Resource Exploitation," *Columbia Journal of World Business* (March/April 1967), p. 26.
34. *Ibid.*
35. United Nations Declaration, "The Establishment of a New International Economic Order," adopted by the Sixth Special Session of the General Assembly, May 1, 1974.
36. "Permanent Sovereignty over Natural Resources" (A/9716/Corr. 1), United Nations, Nov. 5, 1974, p. 4.
37. *New York Times,* January 18, 1975, p. A1.
38. Freyman, "Mineral Resources and Economic Growth," p. 23.
39. Edward Miles, "The Dynamics of Global Ocean Politics," in Douglas

Johnston, ed., *Marine Policy and the Coastal Community: Studies in the Social Sciences* (London: Croom Held, 1976), p. 22.

40. John G. Laylin, "The Law to Govern Deepsea Mining Until Superseded by International Agreement," *The San Diego Law Review* (May 1973), p. 440.
41. H. E. Goeller and A. M. Weinberg, "The Age of Substitutability," *Science*, 191, 4227, (Feb. 20, 1976), 683–88.

Chapter 5

Interests in Policy Formation

Just as the deep seabed issue has been greatly affected by the resource politics described in Chapter Four, so too has this issue been formed by the specific context of UNCLOS negotiations and the pluralism of interests in the United States concerning sea law.

It will be shown in this chapter that negotiation of the seabed issue within UNCLOS, under the terms agreed to in the early 1970s, has tended to prolong the negotiations and to discourage nations from taking matters into their own hands. It will also be seen that the many and varying ocean interests of the United States have prevented the government from throwing its complete support behind the domestic nodule industry. The internal struggle within the U.S. government to formulate a coherent and widely accepted ocean policy is discussed, as is the divergence of administration policy and that expressed by segments of the Congress.

The politicization of the deep seabed issue can be traced to Ambassador Pardo's 1967 speech and the subsequent convening of the U.N. Seabed Committee. Before this time potential deep-sea miners generally operated in a policy vacuum. All potential miners recognized the legal ambiguities involved in mining beyond national jurisdictions, but they could hardly have forecast the participation of the entire international community in resolving them. The year 1967, therefore, marked a turning point, whereby lawyers and politicians slowly but steadily gained ascendency in determining the future development of the entire enterprise.

Mining organizations have traditionally eschewed the political spotlight, hoping to be left alone in their chosen tasks. Prospective miners of manganese nodules have not been so fortunate. During 1968 and 1969, the U.N. Ad Hoc Seabed Committee consisted of forty-two national delegates. By 1972, the permanent Seabed Committee totaled ninety-one members. Finally, at the first substantive UNCLOS meeting in 1974, the total number of states participating reached over 140, representing the largest multilateral conference that had ever been assembled. The bulk of these states were relatively new, developing nations. As a matter of reference, at the 1958 Law of the Sea Conference, fifty developing nations participated, and by 1974 the number had increased to 106.

UNCLOS Proceedings

To put it mildly, mining officials have resented the intrusion of international politics into what they would prefer to perceive as a purely technical endeavor. We have already indicated that there is considerable risk and technical uncertainty associated with the enterprise, risks that the miner normally accepts and with which he is familiar. Politics, however, adds to and compounds severalfold the miner's gamble. The inconclusive nature of the negotiations, which have been drawn out over a decade, has simply added to the antagonism between miner and politician. Reflecting the frustration of miners, the *Wall Street Journal* as early as 1974 characterized UNCLOS negotiations as simply existing to provide the international set a chance to travel from "spa to shining spa."[1] In much the same vein, Deepsea Ventures executives have written,

> Odysseus was appointed the task of walking inland with a ship's oar over his shoulder until he found inhabitants so ignorant of the sea as to mistake the oar for a threshing tool. At this point he was to make sacrifice to Poseiden, God of the Sea, in the hope that this act would calm the godly anger and moderate the winds and waves at sea. Today, some three thousand years later, descendants of these inhabitants, still ignorant of the ways of the sea and the men who sail thereon, are reversing the journey to troop to the shores of the East River; there to generate great winds and waves in an attempt to inhibit the acquisition of knowledge and values from the sea.[2]

The attempt to deprecate the efforts of the negotiators at UNCLOS stems from the desire by miners to reduce the perceived legitimacy of the international undertaking. Fearful that they will be unable to obtain their desired goals from international negotiators, miners have hoped to achieve them by bypassing the negotiations altogether. Lip service has been paid over the years to the desirability of concluding an international accord, but miners recognized very early in the negotiations that the chances of obtaining the "right" kind of accord were slim. Early in the 1970s, mining officials were already claiming that the seabed forum "had lost touch with reality."[3] Hoping to gain government support for "interim" commercial mining, Dubs stated the industry predicament as follows: "Industry has progressed to the point where the next steps require very large development expenditures and they are deeply concerned about making such expenditures, since their investment may be negotiated away at a Law of the Sea Conference and there is no assurance that their own government will afford them protection against interference in the interim period before a treaty. Without a stable and predictable investment climate, progres toward commercial recovery will be halted or reduced to an insignificant rate."[4]

Much of the problem and delay stems from the UNCLOS agenda, drawn up in 1972-73 by the U.N. Seabed Committee. At the insistence of several developing nations, particularly the Latin Americans, the agenda was set to include virtually every ocean issue that might conceivably be disputed. Rather than produce the kind of conference that Pardo had advocated in 1967–one that would deal exclusively with the creation of an international agency to regulate deep seabed mining–it was decided to open up the conference to virtually all significantly divisive ocean issues. Observers recognized immediately that this could have a significant impact on the deep seabed mining issue, as the possibility for tradeoff among issues was obvious. What was not so keenly appreciated was the absolutely crucial decision to treat these issues as a "package," i.e., an integrated whole which would be both negotiated simultaneously and resolved simultaneously. In other words, there was a commitment from the beginning, not to negotiate or put in treaty form "pieces" of negotiations, but instead to wait until all the "pieces" had fallen into place. As we will see in this and subsequent chapters, this decision has had a profound effect upon the resolution (or lack thereof) of the seabed mining controversy.

To appreciate the ambitious scope of UNCLOS proceedings, it may be useful to set forth the agenda. UNCLOS has from the beginning been split into three main committees: The first has dealt with the seabed issue; the second has considered general boundary delineation, navigation (both commercial and military), and the rights of land-locked or shelf-locked states; the third has covered the protection of the marine environment, scientific research, and the transfer of technology. Although not envisioned in the early years of seabed negotiations, a fourth committee has been formed to deal specifically with the settlement of marine disputes. The following is UNCLOS's initial formal agenda:

ALLOCATION OF AGENDA ITEMS BETWEEN COMMITTEES

All Main Committees

Items to be dealt with by each main committee insofar as they are relevant to their mandates

Item 1. Regional arrangements
Item 2. Responsibility and liability for damage resulting from the use of the marine environment
Item 3. Settlement of disputes
Item 4. Peaceful uses of the ocean space; zones of peace and security

First Committee

Items to be considered by the first committee

Item 5. International regime for the seabed and ocean floor beyond national jurisdiction
- 5.1 Nature and characteristics
- 5.2 International machinery: structure, functions, powers
- 5.3 Economic implications

5.4 Equitable sharing of benefits, bearing in mind the special interests and needs of the developing countries, whether coastal or land-locked

5.5 Definition and limits of the area

5.6 Use exclusively for peaceful purposes

Item 6. Archaeological and historical treasures on the seabed and ocean floor beyond the limits of national jurisdiction

Second Committee

Items to be considered by the second committee

Item 7. Territorial sea

7.1 Nature and characteristics, including the question of the unity or plurality of regimes in the territorial sea

7.2 Historic waters

7.3 Limits

7.3.1 Question of the delimitation of the territorial sea; various aspects involved

7.3.2 Breadth of the territorial sea; global or regional criteria; open seas and oceans, semienclosed seas and enclosed seas

7.4 Innocent passage in the territorial sea

7.5 Freedom of navigation and overflight resulting from the question of plurality of regimes in the territorial sea

Item 8. Contiguous Zone

8.1 Nature and characteristics

8.2 Limits

8.3 Rights of coastal states with regard to national security, customs and fiscal control, sanitation, and immigration regulations

Item 9. Straits used for international navigation

9.1 Innocent passage

9.2 Other related matters, including the question of the right of transit

Item 10. Continental shelf

10.1 Nature and scope of the sovereign rights of coastal states over the continental shelf; duties of states

10.2 Outer limit of the continental shelf: applicable criteria

10.3 Question of the delimitation between states; various aspects involved

10.4 Natural resources of the continental shelf

10.5 Regime for waters superjacent to the continental shelf

10.6 Scientific research

Item 11. Exclusive economic zone beyond the territorial sea

11.1 Nature and characteristics, including rights and jurisdiction of coastal states in relation to resources, pollution control, and scientific research in the zone; duties of states

11.2 Resources of the zone
11.3 Freedom of navigation and overflight
11.4 Regional arrangements
11.5 Limits: applicable criteria
11.6 Fisheries
  11.6.1 Exclusive fishery zone
  11.6.2 Preferential rights of coastal states
  11.6.3 Management and conservation
  11.6.4 Protection of coastal states' fisheries in enclosed and semienclosed seas
  11.6.5 Regime of islands under foreign domination and control in relation to zones of exclusive fishing jurisdiction
11.7 Seabed within national jurisdiction
  11.7.1 Nature and characteristics
  11.7.2 Delineation between adjacent and opposite states
  11.7.3 Sovereign rights over natural resources
  11.7.4 Limits: applicable criteria
11.8 Prevention and control of pollution and other hazards to the marine environment
  11.8.1 Rights and responsibilities of coastal states
11.9 Scientific research

Item 12. Coastal state preferential rights or other nonexclusive jurisdiction over resources beyond the territorial sea
12.1 Nature, scope, and characteristics
12.2 Seabed resources
12.3 Fisheries
12.4 Prevention and control of pollution and other hazards to the marine environment
12.5 International cooperation in the study and rational exploitation of marine resources
12.6 Settlement of disputes
12.7 Other rights and obligations

Item 13. High seas
13.1 Nature and characteristics
13.2 Rights and duties of states
13.3 Question of the freedoms of the high seas and their regulation
13.4 Management and conservation of living resources
13.5 Slavery, piracy, and drugs
13.6 Hot pursuit

Item 14. Land-locked countries
14.1 General principles of the law of the sea concerning the land-locked countries
14.2 Rights and interests of land-locked countries
  14.2.1 Free access to and from the sea: freedom of transit, means and facilities for transport and communications

14.2.2 Equality of treatment in the ports of transit states
14.2.3 Free access to the international seabed area beyond national jurisdiction
14.2.4 Participation in the international regime, including the machinery and the equitable sharing in the benefits of the area
14.3 Particular interests and needs of developing land-locked countries in the international regime
14.4 Rights and interests of land-locked countries in regard to living resources of the sea

Item 15. Rights and interests of shelf-locked states and states with narrow shelves or short coastlines
15.1 International regime
15.2 Fisheries
15.3 Special interests and needs of developing shelf-locked states and states with narrow shelves or short coastlines
15.4 Free access to and from the high seas

Item 16. Rights and interests of states with broad shelves
Item 17. Archipelagos
Item 18. Enclosed and semienclosed seas
Item 19. Artificial islands and installations
Item 20. Regime of islands
20.1 Islands under colonial dependence or foreign domination or control
20.2 Other related matters

Item 21. Transmission from the high seas

Third Committee

Items to be considered by the third committee

Item 22. Preservation of the marine environment
22.1 Sources of pollution and other hazards and measures to combat them
22.2 Measures to preserve the ecological balance of the marine environment
22.3 Responsibility and liability for damage to the marine environment and to the coastal state
22.4 Rights and duties of coastal states
22.5 International cooperation

Item 23. Scientific research
23.1 Nature, characteristics, and objectives of scientific research of the oceans
23.2 Access to scientific information
23.3 International cooperation

Item 24. Development and transfer of technology
24.1 Development of technological capabilities of developing countries

24.1.1 Sharing of knowledge and technology between developed and developing countries
24.1.2 Training of personnel from developing countries
24.1.3 Transfer of technology to developing countries

The vast scope of UNCLOS proceedings, evident by the agenda, placed the seabed issue within a very broad panorama. UNCLOS became the concern, not only of prospective manganese nodule miners, but also other special interests having their own respective stakes in the outcome. With UNCLOS encompassing virtually every ocean activity, fishermen, oil and gas interests, the Defense Department, and scientists all recognized that their traditional ocean activities could be greatly affected. For this reason, they lobbied the U.S. government long and hard during the formative years of U.N. Seabed Committee deliberations to ensure that their respective interests would be reflected in "appropriate" U.S. policy.

The tortuous bureaucratic infighting involving special interests and various government agencies in the formulation of U.S. ocean policy has been described elsewhere.[5] A lack of continuous high-level administrative involvement has been a prominant characteristic of the government's formation of ocean policy, and lower-level bureaucrats have made the primary decisions. High-level administrative officials have generally been called in only to resolve disputes that arise among the lower-level bureaucrats. The two main reasons for high officials delegating their policy-making responsibilities are, first, that they perceive issues other than ocean matters as more important and hence more worthy of their full attention; and second, that the legal, political, and economic complexities of ocean issues are overwhelming, given the time constraints within which these officials operate.[6]

Recognizing that the United States might not be able to obtain all it wanted in UNCLOS negotiations and fearful that some national goals would thereby have to be sacrificed, all special-interest groups made a sustained attempt to convince policy-makers that their respective concerns were crucial to the nation. The lack of a clear, high-level arbitrator in these matters made the infighting all the more intense. In comparison to the lobbies for fish, oil, and gas, and the Department of Defense, the small, prospective manganese nodule industry was clearly no match. This fact led to a concern on the part of miners, which continues to exist, that their interests would be traded off to obtain the more obvious and substantial benefits associated with other ocean activities. Time and again administration officials have sought to reassure the miners that their interests would not be sacrificed as part of a bargain to obtain the desired UNCLOS treaty, and to this day the administration has remained faithful to this promise.

On the other hand, the administration has not been willing to sanction the unilateral licensing of domestic firms for commercial mining, as the miners have requested. It is the author's belief that had UNCLOS negotiations been confined solely to seabed mining, the government would have by this time licensed seabed miners, at least on an interim basis. However, because UNCLOS involves a vast number of important ocean issues, U.S. officials have been extremely reluctant to take any action that might jeopardize their successful negotiation.

The fact is that outside of the seabed mining issue (and the proposed regime for scientific research), the United States has generally been pleased with UNCLOS progress—as reflected in the various negotiating texts that have emanated from annual sessions. There finally appears to be general international consensus on boundaries, which would, if enacted in treaty form, eliminate the many conflicting claims—and thereby significantly reduce the chances for international conflict based upon legal ambiguities, e.g., the "cod war" and the "tuna war." International consensus on a 12-mile territorial sea and a 200-mile exclusive economic zone is surely within the U.S. national interests, given our exceedingly long coastal area. General agreement at UNCLOS, also within the U.S. interest, has been achieved for military and commercial passage on the high seas and through straits, pollution standards and responsibilities, and binding procedures for the settlement of disputes. Were the U.S. government to take unilateral action on the deep seabed, therefore, the above agreements, which were painstakingly put together over years, could all come undone. Thus, the "package" nature of UNCLOS negotiations has proved a serious restraint upon U.S. government action on the seabed issue.

U.S. Administration

Since its initial commitment to negotiating the deep seabed issue within the UNCLOS format, the U.S. administration has steadfastly reaffirmed its pledge to produce an acceptable legal regime through this forum. As was just stated, much of this devotion to UNCLOS results from the interlocking nature of the various issues before it. It also results from the belief that essentially the international community has little choice but to accept an international seabed regime well within the U.S. interests. This perception stems from the fact that it is the industrial nations, and particularly the United States, that contain the organizations having the actual operating technology and capability to mine for manganese nodules. Without the acquiescence of these nations there would clearly be no functioning mining enterprise. In congressional testimony the chief U.S. negotiator on the deep seabed issue stated this view precisely: "I think most of the developing countries realize that without the highly industrialized countries being accommodated in this negotiation, it is not likely that there will be a treaty which will be of any use to the developing countries.[7]

The view that the United States is dealing from a position of strength, and the realization that mining interests are carefully observing their every move, have greatly influenced the negotiating stance of the U.S. team at UNCLOS. Contrary to the perception of some observers,[8] the U.S. position throughout the negotiations has remained, in essence, rather inflexible. (This topic will be addressed in detail when UNCLOS negotiations are examined in Chapter Nine.) What the government has sought throughout UNCLOS negotiations is a legitimization of its preferred future. Developing nations have been equally inflexible in putting forth their vision of a preferred international regime. What has resulted at UNCLOS, therefore, is not a real negotiation involving sufficient compromises of interest, but a stalemate in which parties are talking past one another. The unwillingness of the United States to bypass the UNCLOS forum in developing a new legal regime, for the reasons alluded to earlier, and at the same time, the unwillingness of U.S. negotiators to make significant concessions, represent the essential dilemma facing policy-makers today. It is evident that something must give, but it is not clear what.

Mining officials have worried from the beginning of the U.N. Seabed Committee debates that their interests would be sacrificed to achieve an international accord. The earliest indication of what might constitute a new legal regime, not only for the deep seabed but throughout ocean space, came in the Draft 1970 U.S. Convention on the International Seabed Area presented for the consideration of the U.N. Seabed Committee.[9]

The United States in 1970 proposed that national jurisdiction over seabed resources be limited to areas shoreward of the 200-meter isobath. Beyond this limit and extending to the outer edge of the continental margin would be an area called an "international trusteeship zone." This area would be under international jurisdiction, and revenues from the exploitation of resources would be shared by coastal nations and the international community (funds would be turned over to an international development agency for assistance to developing countries). Beyond the continental margins, all seabed exploitation would be regulated by an international agency, and a share of the revenues, again, would go into a fund to be distributed to developing nations. This agency would be controlled primarily by industrial powers and would license mining companies. In short, the proposal presented an interesting mix of national and international jurisdictions. Since national jurisdiction over the continental shelf was firmly established by 1970 and could extend beyond the 200-meter isobath (because of the 1958 U.N. Continental Shelf Convention), the U.S. proposal was actually a call for relinquishing national sovereignty over potential resources.

This proposal never did gain international support, which prevented what could have been a long and divisive squabble in this country over its merits. What was apparent from the plan was that the U.S. government, or at least the Departments of State and Defense, were willing to forego national claims over vast expanses of the seabed in order to secure an international treaty. The Defense Department was particularly interested in narrow national boundaries as it wanted to ensure freedom of navigation over as much of the oceans as possible. As Hollick has recounted, oil interests were disturbed with the plan, since they wanted no part of any new and amorphous international regime.[10] Although manganese nodule miners, unlike the oil interests, were not unprepared for working under an international body—and the specific terms of the arrangement were certainly favorable from their vantage point—they recognized the overall plan as basically a trade of U.S. economic interests for securing its diplomatic and strategic interests. It was their fear that the negotiating team would move even further in this direction when negotiations actually began.

We have already seen that government support of deep seabed mining as a high-seas freedom is not judged by mining companies to be a sufficient guarantee of mining rights. The refusal of the State Department to grant Deepsea Ventures "diplomatic protection" in 1974 was perceived as another indication of the government's coldness toward the mining enterprise. In addition, industrial officials in numerous congressional hearings have referred to the direct government assistance that other nations are providing their developing nodule industries.[11] The governments of Great Britain, France, Germany, and Japan, among others, have come forth with substantial capital funds to assist their domestic miners in developing the necessary technology; in contrast, the U.S. government has contributed little. Manganese nodule mining is, of course, not a special case in this regard. Traditionally, the line is drawn fairly carefully in this country as to where public and private responsibilities begin and end. In other countries the demarcation between public and private sectors is not so precise. This is particularly true in Japan, where government and industry work hand in hand, especially concerning mineral exploitation, as that small, resource-poor island places a high priority on obtaining the necessary minerals to support industrialization. Craven, in a multinational study, characterizes the Japanese system as the ideal manager of ocean resources. The United States, however, with its careful delineation of proper public and private roles and its strong antitrust provisions, serves as the antithesis to the Japanese system.[12] The fact, therefore, that potential competitors to U.S. mining ventures are being substantially rewarded by their respective governments is a constant source of irritation to U.S. miners, who have to progress without government largess.

However, as has been pointed out by others, U.S. mining companies often take their position regarding government neglect to an extreme.[13] The government has certainly subsidized valuable research; without the government-supported research at such institutions as Lamont-Doherty Geological Observatory and Scripps Institution of Oceanography, U.S. miners would have had to carry on much preliminary analysis under their own auspices and at their own expense. In addition, companies have been able to write off a percentage of their capital expenses on their corporate income taxes, which represents a clear subsidy to industry, even though it is not as direct and obvious as an outright grant. Hence, although it appears as if other nations have assisted miners more generously, one should not be blind to the assistance that has been provided by the government.

It was, however, the Treasury Department, which entered the debate on ocean policy rather late (1973–74), that objected most strenuously to the terms of the 1970 U.S. draft. Recognizing that it served the purposes of the Departments of Defense and State (i.e., through its search for what some have derogatorily referred to as "the perfect treaty") but represented considerable economic sacrifice, the Treasury Department strongly opposed the draft. Through their efforts within the delegation, in addition to the pressure applied by the oil industry, the government cautiously backed off the proposal, until by 1974, at the initiation of the UNCLOS Caracas session, the chief representative indicated the U.S. government's acquiescence in the establishment of 200-mile exclusive economic zones. This policy, of course, was completely contrary to the support of narrow national extensions proposed in the 1970 draft. The Treasury Department has been equally adamant concerning any talk of economic sacrifice on the deep seabed issue, but it has never abandoned the administration position that unilateral legislation allowing companies to be licensed directly would be both premature and unwise in light of UNCLOS proceedings. In fact, as we will see later in this chapter, the Treasury Department has been most forceful in its opposition to unilateral initiatives.

In short, the U.S. administration throughout the 1970s has never swayed from its original goal of achieving an international accord on seabed mining. Session after session has passed with scarcely any progress being made, and yet the administration persists. As stated previously, had only seabed mining been taken up at UNCLOS, negotiations probably would have been terminated long ago, but the administration has not wanted to jeopardize the progress that has been made in other areas. Hence, whereas the miner's interests, which are far more narrow and focused than the administration's, are patiently reasserted by the administration before every UNCLOS session, the patience of the prospective seabed miner for progress has long been exhausted.

U.S. Congress

The miner's plea for immediate action, however, has not been completely neglected within the U.S. government. Miners have found pockets of strong support within Congress.

Possible federal legislation that would license domestic miners before an UNCLOS treaty—and in clear violation of the U.N. Moratorium and Declaration of Principles Resolutions—was introduced into Congress as early as 1971 by Senator Lee Metcalf (D., Montana). As Chairman of the Senate Subcommittee on Minerals, Materials, and Fuels (of the Senate Committee on Interior and Insular Affairs), he first held hearings on the bill (S. 2801) in 1972. Some of the subcommittee's staff had attended deliberations of the U.N. Seabed Committee in the late 1960s and early 1970s and were alarmed at the options for international control that were being considered.[14] Consequently, they prevailed upon Metcalf to offer a practical alternative to what might come out of UNCLOS, one which they felt would better ensure the national interest. It should also be noted that Congressman Thomas Downing (D., Virginia) introduced a bill identical to Metcalf's in the House in early 1972 (H.R. 9). Not coincidentally, the corporate home of Deepsea Ventures was located in the district Downing represented. Despite or because of this fact, it was Metcalf and his staff that took the lead in pressing for action on the bill.

Senator Metcalf was completely frank about the origins of S. 2801, which was drafted by the American Mining Congress (within this organization, it received strong input from Kennecott Copper and Deepsea Ventures). He claimed that he introduced it for "discussion purposes only," to provide the mining companies a chance to tell their side of the story. Under S. 2801, qualified mining entities would be able to obtain from the Secretary of Interior licenses for blocks of seabed 40,000 square kilometers (which approximates the size of West Virginia). The mining entity would control the entire block or concession for up to ten years, during which time extensive exploration would take place. After ten years, and presumably when commercial mining would commence, the mining entity would be forced to relinquish 30,000 square kilometers of the claim. Hence, the first ten years would be used to identify the richest deposits within the assigned block, and subsequently, much of the area of little commercial interest would be forfeited.

The potential licensing of such vast areas of the seabed raised considerable controversy. Even Metcalf claimed that the 1870 corporate giants would never have dared ask for so much land.[15] The Federation of American Scientists attacked the bill, saying that conflict could well arise if the oceans were "carved up on a first-come, first-served basis, as was colonial Africa in the 19th century";[16] and John Mero, the original manganese nodule entrepreneur, stated, "The grab that Kennecott Copper and Deepsea Ventures are planning with the apparent innocent support of the American Mining Corporation is truly monumental."[17]

In defense of their proposal, mining representatives pointed out that unlike conventional mining deposits, which are vertical in nature, manganese nodule deposits rest on a horizontal plane. Vast areas of the seabed were required, therefore, to make up for the lack of a vertical dimension. Moreover, they emphasized that such mining was only an *interim* approach, and that once the United States had signed a satisfactory UNCLOS treaty, their mining operations would be subsumed within the accepted international framework.

The claim that domestic enabling legislation would only be an interim measure until a more definitive international approach is reached has never been entirely convincing. Interim proposals have a way of becoming more than temporary over time. It is clear that if commercial mining were underway it would be extremely difficult for U.S. senators to ratify a treaty that might seriously disrupt it. Having a "foot in the door" is a recognized means of influencing possible future events. The "interim" label has always been stamped on subsequent bills during the 1970s, however, to make the measure more palatable to skeptical senators and representatives.

Some of the other features of the early bill deserve elaboration. To obtain blocks of 40,000 square kilometers, qualified miners would have to pay a fee of $5,000. In order to fulfill what they perceived to be the dictates of the principle of the common heritage of mankind, mining companies would agree to establish an escrow fund, setting aside an undefined percentage of their income tax and license fees to directly contribute to the welfare of less developed countries. Also under the bill, the United States would recognize reciprocal claims by the mining companies of other nations to areas of the seabed. It was felt, therefore, that once the United States took the lead in unilateral mining, other nations would follow suit and enact similar legislation. How one would reconcile differing mining laws and jurisdictions, or conflicting claims to specific areas of the seabed, was not touched upon. Finally, some sort of investment protection would be an integral part of the legislation; i.e., miners would be reimbursed if their integration into an international regime could only be accomplished at a loss to the company.

The issue of investment protection and guarantees by the government has come to center stage in the debate over interim legislation. Mining officials claim that without some form of government guarantees against losses from third-party interference or from their eventual integration within an international accord produced at UNCLOS, mining companies cannot obtain the necessary risk capital. This claim was confirmed by the testimony of Chase Manhattan Bank Vice-President C. Thomas Houseman (as cited in Chapter One). The most immediate need to domestic miners, therefore, is to obtain the funds necessary to advance toward commercialization. Despite the willingness of some congressmen, such as Metcalf, to provide them with the backing necessary, the measure has not yet been passed as legislation.

For many years, the administration testified against passage of a bill that would permit licensing of domestic mining companies, claiming that such action was premature and would be harmful to the satisfactory achievement of an UNCLOS treaty. After each inconclusive UNCLOS session, administration spokesmen would be called to testify before Senator Metcalf or other congressional representatives, and to indicate whether the lack of positive results at UNCLOS had in any way altered administration opposition to interim seabed legislation. Invariably, administration officials cited progress being made at UNCLOS sessions which led them to resist supporting unilateral initiatives. Up until the fall of 1977, therefore, the administration was actively involved in two separate fronts; first, attempting to reach agreement at UNCLOS within the international community, and second, trying to hold off Congress from taking what it perceived as precipitous action. In the fall of 1977, however, Ambassador Richardson, chief negotiator for the United States at UNCLOS, recognizing that he was not having much success on either front, reversed administration policy and announced support for interim legislation. As we will see in the following discussion of seabed bills, however, the type of legislation favored by the administration differed substantially from that favored by Senator Metcalf.

Seabed Mining Legislation

Although it is claimed that the 1974 seabed mining bill, S. 1134, was the product of congressional staff[18] and not simply the American Mining Congress, very little of substance was changed from the earliest bill introduced by Senator Metcalf. The bill S. 713, considered in 1975 and 1976, stated that license fees to obtain blocks of 60,000 square kilometers of seabed would only amount to the administrative and processing costs associated with the application. After fifteen years, or when commencement of commercial mining would occur, miners would have to give up 50 percent of the block. In addition, fairly detailed work requirements were specified.

The most noteworthy aspect of S. 713, however, was its explicit provision for an investment guarantee and an investment insurance program. Specifically, it declared that licensed companies would be entitled to indemnification if their investments were reduced in value by the accession of an international treaty. Further, an investment guarantee and insurance fund would be established, whereby companies for an undetermined premium could receive compensation for existing investment and "future value and profits" if their activities were interfered with or prevented "by action of any person against whom a legal remedy may not exist." When pressed to put a figure on how much such compensation might come to, Metcalf and others conceded that they had no way to estimate the possible U.S. government liability.[19] Presumably, though, it could run into billions of dollars.

Ambassador Richardson in 1977 made it clear that the administration would only support legislation that (1) designated no specific sites for mining, (2) spelled out the regulatory details for exploration but not exploitation, and (3) eliminated all references to government investment or loan guarantees. Hence, the switch of the administration to support for legislation did not mean support for the legislation desired by mining companies.

Administration opposition to investment guarantees first became evident in the mid-1970s. In speaking before the Metcalf subcommittee, Deputy Assistant Secretary of the Treasury for Trade and Raw Materials Policy J. Robert Vastine identified the areas of the bill to which the Treasury Department strongly objected, namely, the investment guarantees and insurance program.[20] In short, he claimed that these two portions of S. 713 effectively shifted the normal risks of business from the companies to the government and that such a shift was unacceptable. His specific points were as follows:

1. The U.S. administration was attempting to produce a common investment policy across broad industry groups that would not involve government handouts. "Treasury believes there is no need to change policy at this time and give special incentives to a limited segment of the non-fuel minerals industry."

2. The passage of such legislation might set a precedent whereby all firms would expect the government to protect them from such political risks.
3. The prospect of paying off companies as a prelude to signing an UNCLOS treaty was unacceptable.
4. The bill would place on the government a liability of undetermined amount.

In short, despite the policy change, one still finds a good deal of reticence within the administration for seabed legislation. Some people within the Defense Department object because such legislation could jeopardize the interests they see being fulfilled in UNCLOS proceedings; some State Department officials object because the legislation might unravel all the delicate compromises negotiated throughout the 1970s; and finally, the Treasury Department objects because it views possible legislation as a very large and unwarranted subsidy to the small nodule industry. The fact that Ambassador Richardson was in Europe with other duties during the last week of the 95th Congress (1978), and not up on Congressional Hill pressing for seabed legislation, is indicative of the degree of administration support for such legislation at this time.

The bedrock of support for interim seabed legislation has been and remains in Congress. With the death in 1978 of Senator Metcalf, and with the loss of key staff members, the focus of mineral industry support has shifted from the Senate to the House. The House Merchant Marine and Fisheries Committee Chairman John Murphy (D., New York) and the Oceanography Subcommittee Chairman John Breaux (D., Louisiana) have been outspoken supporters of domestic seabed miners. Murphy introduced in 1976 a bill for unilateral seabed licensing similar to the Metcalf bill, and reintroduced it (commonly referred to as the Murphy/Breaux bill, H.R. 3350) in the 1977 and 1978 sessions. Both Murphy and Breaux are staunchly committed to seeing that domestic mining groups are able to mine as soon as possible.

Speaking in 1976, Murphy stated, "Congress can no longer sit back and watch this erosion of our technological lead. We can no longer sit back and watch the State Department bargain away U.S. interests. We can no longer sit idly and watch as a secure source of minerals evaporates before our eyes. It is time to act. We must enact legislation into law to enable the U.S. ocean mining industry to proceed with the development of their technology and the recovery of the manganese nodules."[21]

References to maintaining our technological lead in manganese nodule commercial development are commonly heard at congressional hearings. Industry spokesmen have often claimed that U.S. corporations are approximately two or three years ahead of foreign firms in the development of technology, and that with the greater support other governments are tendering their national firms, the U.S. lead is eroding. Despite the fact that no explanation is ever given for why there need be any "race" to the deep seabed–particularly in light of the abundance of first-generation sites—this argument must serve some utility since it is repeated so often.

It would be a mistake, however, to give the impression that Congress speaks with one voice, and that it universally supports the type of legislation favored by seabed miners. Differences have become clearer as the pressure for some sort of congressional action has mounted. Divergence was evident in 1976 when the Senate Subcommittee on Minerals, Materials, and Fuels favorably reported out S. 713, which also was passed by a unanimous quorum vote of the larger Senate Committee on Interior and Insular Affairs of which the subcommittee is a part. Because of its far-ranging implications, however, the bill, before it could reach the floor for a vote, had to be reported out favorably by the Senate Foreign Relations Committee, the Senate Armed Services Committee, and the Senate Committee on Commerce. In late May and early June 1976, the Foreign Relations Committee voted negatively on the bill, and both the other two committees voted to report the bill without recommendation. Consequently it was never taken up by the full Senate during 1976. The Armed Services Committee stated that it could be persuaded S. 713 was necessary, but it wanted to give UNCLOS negotiators another chance during the forthcoming session.[22] The primary objections of both the Senate Foreign Relations Committee and the Senate Commerce Committee were similar to that of the Treasury Department and focused on the investment guarantee and insurance plan. The report of the Senate Commerce Committee stated,

> Section 13 of the bill would create an unprecedented requirement that the United States indemnify a company for the diminution of its investment resulting from an international agreement signed and ratified by the United States. It is questionable whether governmental action of this kind falls within the protections of the Fifth Amendment to the U.S. Constitution requiring that just compensation be paid for any taking of private property for public use.[23]

For those senators and congressmen, therefore, who had long given up on UNCLOS negotiations, there were still the investment guarantee and insurance provisions which made domestic licensing unpalatable. Although Metcalf could persuade many that UNCLOS was nothing more than a futile exercise, he could not present a sufficiently strong enough case that "open-ended" subsidies were a necessary component of mining.

Divisions within Congress became more pronounced in 1978. In the House, three committees (i.e., Merchant Marine and Fisheries, Interior, and International Relations) reported out competing versions of seabed legislation. The major issues of contention were (1) international revenue-sharing provisions, (2) investment guarantees, (3) lead agency designation, and (4) U.S. vessel requirements. In addition to these three committees, the House Ways and Means Committee argued for the deletion of any international revenue-sharing provisions.

Because of administration pressure and the opposition of the House Interior and International Relations Committees, the Murphy/Breaux bill was amended to delete the investment guarantee provisions (which originally called for investment guarantees against losses of $350 million per mine site or 90 percent of a mining firm's investment—whichever was less). The various committees could not agree on which agency—either the Commerce or the Interior Departments—should regulate seabed mining.

On July 26, 1978, the issues were resolved by House passage of a compromise bill, H.R. 12988, by a vote of 312 to 80. The bill deleted specific investment guarantees but included a "declaration of Congressional intent," to the effect that U.S. UNCLOS negotiators are instructed to obtain treaty provisions that would allow companies that have started mining to continue their operations under substantially the same terms, conditions, and restrictions. Given this provision, some observers believe that the companies would have a firm legal basis to obtain government compensation should an UNCLOS treaty disrupt their operations. The bill also provides for .75 percent of the gross receipts from mining to go into an escrow fund to be handed over to the Authority when integration within an international treaty is consummated. Congressman Murphy, pushing U.S. merchant marine interests, succeeded in getting into the bill a requirement that ships used for mining and processing of manganese nodules be registered under the U.S. flag; and in a close vote the Commerce Department was chosen as the lead regulatory agency.

After House passage, the Senate took up the mining bill in earnest. Just as in the House, the preferred industry bill, sponsored now by Senators Jackson and Bumpers of the Energy Committee, encountered considerable opposition. In this case, it was the Senate Commerce, Foreign Relations, and Finance Committees that took issue with parts of the industry bill. The inability of these committees to reconcile their differences, and the objections of such senators as Abourezk, Kennedy, and Muskie, prevented the measure from reaching the Senate floor in 1978. Hence, the United States came closer to unilateral legislation in 1978 than it had in the past, but barely fell short of the mark.

Conclusion

Despite the continual threats of various congressmen since 1971 that the United States is prepared to move ahead unilaterally on the seabed issue, these threats have yet to be carried out. Realization that there is a large segment of Congress anxious to see mining development proceed has been a constant backdrop to UNCLOS proceedings; and yet this realization has not apparently softened the negotiating stances of UNCLOS delegates from developing nations. It is the author's belief that should unilateral legislation pass in 1979, it will again fall short of the presumed intended goal, namely, to facilitate agreement at UNCLOS.

The U.S. seabed negotiator stated in 1975, "Events are overtaking us. Technology that didn't exist when we started on this in 1968 has developed while we continue to talk."[24] The idea that technological development would, in effect, preempt UNCLOS negotiations and make the debate academic has not been widely accepted at UNCLOS. In fact, both the U.S. administration and the UNCLOS forum have up to this point successfully resisted an approach based on narrow technical determinism. Unfortunately, united opposition to the pleas from mining companies for haste does not constitute a sufficient condition for reaching international accommodation.

Notes

1. *Wall Street Journal,* August 29, 1974, p. 7.
2. John E. Flipse and Richard J. Greenwald, "The Marine Operator's Role in the Rational Formulation of Principles of Law Governing Mining Activities in 'Shared' Ocean Space," *Marine Technology,* Conference Proceedings of the Marine Technology Society, Washington, D.C., June 29–July 1, 1970, p. 573.
3. Leigh S. Ratiner and Rebecca L. Wright, "United States Ocean Mineral Resource Interests and the United Nations Conference on the Law of the Sea," *Natural Resources Lawyer* 6, 1 (Winter 1973), 716.
4. *Current Developments in Deep Seabed Mining,* Part 1, Hearings before the Senate Minerals, Materials, and Fuels Subcommittee, 94th Congress, 1st Session, Nov. 7, 1975, p. 3.
5. Ann L. Hollick and Robert E. Osgood, *The New Era of Ocean Politics* (Baltimore: Johns Hopkins University Press, 1974); for a condensed version see Ann L. Hollick, "The Clash of U.S. Interests: How U.S. Policy Evolved," *Marine Technology Society Journal,* 8, 6 (July 1974), 15–28.
6. Hollick, "The Clash of U.S. Interests," p. 16.
7. Statement in *Status Report on Law of the Sea Conference,* Part 4, Hearings before the Senate Subcommittee on Minerals, Materials, and Fuels, 94th Congress, 1st Session, Oct. 29, 1975, p. 1457.
8. Congressman Murphy claims that the State Department has "repeatedly yielded to the Group of 77 in a headlong rush, an almost masochistic effort, to reach a settlement," *Deep Seabed Mining,* Hearings before the House Subcommittee on Oceanography, 94th Congress, 2nd Session, Feb. 1976, p. 79; also see Robert L. Friedheim and William J. Durch, "The International Seabed Resources Agency Negotiation and the New International Economic Order," Presented at the 1976 Annual Meeting of the American Political Science Association, Chicago, Ill., Sept. 2–5, 1976.
9. *Draft United Nations Convention on the International Seabed Area: Working Paper Submitted by the United States of America* (A/AC. 138/25), June 1970.
10. Hollick, "The Clash of U.S. Interests," p. 18.

11. *Mineral Resources of the Deep Seabed,* Part 2, Hearings before the Senate Minerals, Materials, and Fuels Subcommittee, 93rd Congress, 2nd Session, March 3–11, 1974, p. 962.
12. John P. Craven, "Industry/Government Relations in Offshore Resource Development," *Offshore Technology Conference,* 5, 2 (1973), II-941–II-961.
13. See the testimony of Sam Levering, in *Mineral Resources of the Deep Seabed,* Hearings before the Subcommittee on Minerals, Materials and Fuels, 93rd Congress, 2nd Session, March 5, 1974, p. 1103.
14. David Stang, "The Donnybrook Fair of the Oceans," *San Diego Law Review,* 9, 3 (May 1972), 569–607.
15. George C. Wilson, "Battle Stirs over Seabed Mines Bill," *Washington Post,* April 25, 1972, p. A-1.
16. *Ibid.*, p. A-9.
17. John Mero, *The Ocean Mining Report,* 7, 8 (Aug. 1973), 4.
18. Kenneth H. Kolb, "Congress and the Ocean Policy Process," *Ocean Development and International Law,* 3, 3 (1976), 277.
19. "Deep Seabed Hard Minerals Act," Report to the Senate (# 94-754), 94th Congress, 2nd Session, April 14, 1976, p. 27.
20. Testimony of J. Robert Vastine before the Senate Armed Services Committee, May 19, 1976.
21. *Deep Seabed Mining,* p. 80.
22. "Deep Seabed Hard Minerals Act," Report to the Senate (# 94-935), 94th Congress, 2nd Session, June 8, 1976, p. 8.
23. *Ibid.*, p. 11.
24. *The Science, Engineering, Economics and Politics of Ocean Hard Mineral Development,* 4th Annual Sea Grant Lecture and Symposium, MIT Sea Grant Program, Oct. 16, 1975, p. 25.

PART 2

International Negotiations

Introduction

The first part of this book has been devoted to developing an understanding of the character and dimensions of the deep seabed mining dispute. Technological, legal, and political aspects of the enterprise have been discussed in detail, particularly about U.S. involvement, to illustrate how they intersect in the development of public policy. As has been noted throughout the previous chapters, domestic policy formation has been played out in the shadow of international negotiations at UNCLOS. It is now time to turn to these negotiations in some detail, as the UNCLOS forum has been the central element affecting deep seabed technological development for a number of years.

In its most basic sense, negotiations in Committee I of UNCLOS—the committee that has had total responsibility for structuring a new legal regime for the seabed beyond national jurisdictions—have been over who controls the deep seabed, on the behalf of whom, and for what purposes. At a lower level of generality, negotiations have concentrated upon the power, structure, and functioning of the future international organization to regulate exploitation beyond national jurisdictions. Even more explicity, debate has centered on (1) the degree of supranationality with which to endow the international organization, (2) the national makeup of its decision-making organs, and (3) the actual rules or conditions under which exploration and exploitation will be conducted. In the subsequent treatment of the above issues, a host of subissues will be revealed to illustrate the complexity of the debate and the extent to which nations differ. It will become clear that most, but not all, issues of contention are North–South or have–have not (technical capability) disagreements. Were technical capabilities more widely diffused among the international community, such a clear North-South split on the issues would probably not exist. This split, however, is becoming increasingly prominent, not only with regard to the seabed issue but also with regard to virtually all resource-based issues.

In reality, it is difficult to isolate the three major issues of contention because they are interrelated and interlocked in many ways. For the purpose of careful analysis, however, each issue will be treated in separate chapters.

Chapter 6

Supranationality

Chapter Six is devoted to examining the UNCLOS debate over the degree of supranationality with which to endow the International Seabed Authority (ISA). The essence of this debate, which has been the issue foremost in the consideration of UNCLOS delegates, resides in the determination of who should establish the conditions of mining and who should carry out actual exploitation. Just as do nearly all the issues under consideration, the controversy over supranationality breaks down essentially on a North–South basis. This chapter details why the North considers the creation of an ISA with limited functions imperative, and why the South, in contrast, feels a highly supranational, operational ISA must be created.

Since World War II there has been a proliferation of international organizations involved in numerous fields of endeavor. For the most part they have been created to facilitate the activities of nations or nongovernmental entities in a specific functional task or to provide a service heretofore not available. Very seldom have they been endowed with sufficient power to direct or authorize operational activities. International organizations in the ocean arena are, for the most part, representative of this standard form. Kay, in his study of these organizations, comments, "The coordination of national programs and actions . . . [in] . . . a service/facilitating role" is the primary mode of operation.[1]

Perhaps the most crucial and enduring debate in seabed negotiations from their inception in 1968 has been whether to create a standard "service-facilitating" international organization (subsequently referred to as "the Authoity") to regulate seabed activities in the area beyond national jurisdiction (subsequently referred to as "the Area") or to create a much more powerful and autonomous international organization.[2] In general, developing nations have favored the creation of a strong supranational organization having control over every aspect of resource exploitation. The essential rationale behind this policy is the belief that only a strong international organization can assure that mineral exploitation will be undertaken for the benefit of all mankind. A strong organiation, in other words, is necessary to counter the influence and power of both public and private mining entities. Commenting specifically on this preference, Kay has stated, "There is no model of an international organization having anything approximating this range of tasks and skills."[3] Clearly, such plans, if adopted, would establish a new precedent in international organization.

The United States and other major national powers would prefer to see the creation of a rather standard international organization having limited goals and functions. Representatives from these countries stress that there is no compelling need for more than the provision of necessary services, and, moreover, a stronger international organization would actually retard the efficiency of resource exploitation: It would provide an unnecessary bureaucratic barrier to efficient operation at best, and could even function as an adversary to existing mining organizations.

The differences between nations on the matter of supranationality can be viewed more constructively if this concept is examined more carefully. Skolnikoff has provided a typology of regulatory and/or management capacities.[4] Obviously, the more an organization carries out the functions cited below, the greater the extent of its supranationality.

1. *Service:* All international organizations perform various services, e.g., exchange of information, gathering and analysis of data, expert consultation, facilitation of national and international cooperative programs, coordination of activities, and joint planning. Such services help national governments control, assesss, and utilize global technologies and their effects.
2. *Creation and allocation of norms:* This function involves the establishment of operational standards and regulations (normally through a negotiating process among representatives) and the allocation of costs and benefits. Few international organizations accomplish this function without requiring near unanimity among representatives.
3. *Observance of rules and settlement of disputes:* Included within this category are the monitoring and enforcement of standards and regulations, arbitration of disputes, and legal adjudication. There are relatively few examples of these functions being performed by international organizations —particularly enforcement and adjudication—as they are normally reserved to sovereign states.
4. *Operation:* Only an extremely few international organizations have operational functions which involve management responsibility for the exploitation of resources or the operation of a specific technology. They can also include the provision of technical assistance to nations, as well as direct financing.

The split on the question of supranationality, in its most basic terms, relates essentially to how nations perceive or approach anticipated activities on the deep seabed. Their perception is shaped by the capabilities or potential capabilities they possess to actually exploit the resources. From these differing perceptions and capabilities arise differing objectives and responses. For example, those developed nations having advanced mining capabilities tend to perceive the deep seabed arena in terms similar to any traditional hard mining effort. Their concern is to have the ore extracted, processed, and marketed as efficiently as possible in order to sustain the industrial system in their countries. Since they will have mining capability themselves in the near future, they find very little need for an organization to play anything other than a limited, facilitating role—or as Skolnikoff terms it, a service role.

But organizations can be created by the international community for purposes other than facilitating functional efficiency.[5] From the perception of the developing nations (having no ocean mining capabilities), something more is at stake than simply traditional mineral exploration. In this view, the fact that the Area and its resources are considered to be the common heritage of mankind requires a totally different approach to the activities occurring there. Consequently, the goal of functional efficiency may be subordinate to other goals, such as concern for economic and political equity, which are inherent in the practical expression or interpretation of this principle.

Creation of Norms

As noted previously, an international organization's discretion to set rules and regulations covering its own operations is an important regulatory and management function. Its importance in regard to the proposed Authority was not emphasized during the early Seabed Committee debates of the 1960s and 1970s, but since the commencement of the Law of the Sea Conference in 1974, it has played a key role. There has been a gradual movement toward meaningful compromise in this area, but a position completely satisfactory to all parties has yet to be found.

Originally, the Group of 77 endeavored to separate the issue of who may mine from how they might mine. Its position was that all development of rules and regulations should be left to the Authority to decide once it was established. To do otherwise would be to limit its power and effectiveness. It was thought that any attempt to include the conditions of mining into the convention establishing the Authority would prematurely freeze the nature of activities in the Area before the true shape and extent of the total endeavor was clearly perceived. Hence, most developing nations insisted that there be no mention of the conditions of exploration and exploitation in the prospective convention at all.

The views of the United States were at the other end of the spectrum. At Caracas it argued that the conditions of exploration and exploitation should be explicitly provided for, approaching the specificity of a "mining code." The United States envisioned an approach similar to the creation of the International Civil Aviation Organization under the 1944 Chicago Convention of International Civil Aviation, in which technical rules, standards, and practices were provided for in the text of the convention, and even more technical annexes based on these conditions were developed subsequently by the organization.[6] Although other developed nations could not always agree with the United States and among themselves on the nature of the conditions to be established, they nevertheless supported the inclusion of certain conditions in the convention.

The purpose of insisting that detailed technical sections be included in any treaty is straightforward, namely, to provide mining entities with certainty in regard to mining operations. In this way, investment can be made on the basis of explicit conditions placed within the treaty itself. As such, the international organization would be limited in the discretion it could exercise in defining and interpreting the rules. It was felt that only through the inclusion of conditions in the convention could mining enterprises have assurances of the "rules of the game" and that they would not change precipitately. Even more fundamentally, it was felt necessary that the system of access to seabed resources be spelled out in the text itself. As has been mentioned previously, the right of access to these resources has been of primary national and corporate interest throughout the negotiations. Fear that the organization would discriminate against U.S. companies for whatever reason would be alleviated, then, by placing in the treaty itself the conditions necessary to ensure access.

Toward the end of the Caracas session, the G-77 brought forward a text of "Basic Conditions" which they proposed for inclusion in the treaty. The text contained seventeen draft articles of a general nature, providing no technical details.[7] In fact, the text resembled more a general statement of guiding principles than a set of mining conditions. Despite its brevity and lack of detail, it was viewed by all developed nations as a positive (though obviously insufficient) step toward reconciliation.

At the Geneva session, considerable effort was directed toward producing an acceptable set of basic conditions to be included in the convention. As a result, a more detailed set was produced as Annex I to the Single Negotiating Text (SNT).[8] Included within Annex I, entitled "Basic Conditions of General Survey Exploration and Exploitation," were sections detailing to some degree (1) the access to the Area and its resources, (2) the qualifications of applicants for mining and their selection, (3) the rights and obligations of enterprises working under contract, and (4) the rules, regulations, and procedures. Mining companies were still not completely satisfied with the degree of specificity within Annex I, but U.S. negotiators withdrew their insistence upon a definitive set of rules and regulations. Debate at Geneva, therefore, centered more upon the desirability of certain basic conditions than upon the degree of specificity in general.

Negotiations in the 1976 and 1977 New York UNCLOS sessions produced even more technical specificity. An expanded version of Annex I, describing the basic conditions of prospecting, exploration, and exploitation, was included in the Revised Single Negotiating Text (RSNT) and the Informal Composite Negotiating Text (ICNT). Also included was an Annex II (in the RSNT) and an Annex III (in the ICNT) describing the proposed nature and functions of the Enterprise, the operational arm of the Authority. Also included in both the

RSNT and ICNT were detailed sections dealing with the system for settling disputes and revenue-sharing plans.

We can see, then, that as negotiations have proceeded, greater technical detail has been spelled out and presumably will find its way into the treaty. Despite this movement toward accommodation, disagreement persists over how much discretion the international organization ought to have in establishing the conditions of access to seabed resources. As stated in the report issued by committee chairmen at the end of the 1976 New York summer session, "It is doubtful that any delegation supports an automatic assurance of access, since there seems to be general agreement that the Authority will presumably have some degree of discretion in applying the relevant provisions of Annex I. The question is rather the degree of allowable discretion and the manner in which that discretion could be used."[9]

Allocation of Costs and Benefits

It was anticipated economic benefits from deep seabed mining that initially drew the attention of the international community to the ocean arena in 1967. With the dwindling resource base available to the international community, however, the question of allocating costs and benefits, as described earlier, has receded into the background. Most developing nations believe now that the revenues from mining will be insignificant once expenses are subtracted.[10] As a result, the chairman of Committee I has consistently stated that the question of revenue from the Area is definitely subordinate to that of international participation and control. If the extent to which issues are discussed at UNCLOS negotiations is any indication of their importance, then revenue-sharing has surely not been one of the major issues.

Most nations, nevertheless, have continued to support the concept of revenue-sharing from deep seabed mining, even if this support has thus far been expressed only in vague and general terms. There are any number of assessment methods of raising revenue, such as license fees, bonuses, royalties, profit-sharing, and production-sharing. Again, there is dispute over whether the Authority itself will determine the assessment and distribution method or it will be written into the convention, and the dispute is again between the North and South. Mining nations, fearing that organizational discretion over the methods of assessment could lead to excessive financial burdens on mining entities, have sought to have explicit revenue-sharing provisions placed in the treaty. Developing nations, on the other hand, have claimed that the Authority needs the freedom to impose "appropriate" revenue-sharing schemes. Establishing a precise formula before mining has commenced, they claim, is clearly premature.

Under the SNT, the methods of both assessment and distribution of revenues are left unspecified, although the power to determine them are clearly left to the Authority. It was only during the New York UNCLOS sessions that serious thinking about revenue-sharing got underway. As mentioned previously, the RSNT included a special appendix on financial arrangements, in which, instead of a single plan, two basic options were presented. Approach A was favored by the United States and mining companies. Though it concedes that the Authority would have the right to draw up "rules, regulations and procedures for financial arrangements," it nevertheless states that the Authority should be guided by the specific principles therein enumerated. The plan provided for apportionment of the mining companies' "net proceeds" on a specific percentage basis. Although actual figures were left out of the appendix, it is clear that they were expected to be filled in when drawing up the treaty. Approach B, favored by the developing nations, calls on the Authority, rather than the treaty, to apportion on a percentage basis the proceeds from mining. Presumably, these figures or percentages could be subject to negotiation between the Authority and mining companies. Moreover, whereas Approach A called for an apportionment of *net profits,* Approach B favored other revenue-sharing options:

1. a scale of progressive charges in monetary terms against the gross value of processed metals derived from the Area, or
2. a scale of progressive charges in kind on a percentage basis of the processed metals derived from the Area, or
3. a fixed sum of money per standard unit of weight of the processed metals derived from the Area

The ICNT moved closer to the preferences of mining countries in the sense that it called for explicit percentages (again left unspecified) and rates to be included in the treaty rather than left for the Authority to decide. On the other hand, it called for substantial payments to be made to the Authority from (1) an annual fixed charge to mine (2) a production charge of an unspecified percent of the market value or of the amount of the processed metals extracted from the contract area, and (3) a share of net proceeds according to the contractor's rate of return on investment.

In sum, nations still appear a good deal apart with regard to revenue-sharing, and even the text in the ICNT is still labeled as preliminary. Whether this issue, in itself, has the potential to prevent an agreement is open to speculation.

Observance of Rules and Settlement of Disputes

A clear North-South split over observance of rules, and particularly settlement of disputes, has not developed. One must add, however, that these issues have not played a central part in negotiations, and many nations still have not expressed a preference about the appropriate power of the Authority in these matters. In contrast to the previous issues discussed, it is primarily the mining nations that advocate giving the Authority strong powers to resolve disputes.

The reason for this anomaly stems from their desire to gain a good deal of protection for the investments in mining that will be made. An Authority or seabed tribunal with powers of adjudication could go a long way in ensuring the protection mining enterprises appear to require.

The United States was first in proposing an Authority seabed tribunal to settle all disputes arising from exploration and exploitation of the Area's resources. Under a plan of compulsory settlement, decisions reached by the tribunal would be binding on all parties to the dispute. The preferred U.S. position is reflected in Article 37(1) of the RSNT, which states, "Judgements and orders of the Tribunal shall be final and binding. They shall be enforceable in the territories of Members of the Authority in the same way as judgements or orders of the highest court of that Member State."

Not all nations are as enthusiastic about compulsory settlement of disputes as is the United States. Several developing nations would support endowing the Authority with some powers, but they are reluctant to make these powers compulsory. As it is, the RSNT encourages parties to a dispute to attempt to seek a solution first through consultation, negotiation, and conciliation procedures. Provision for the creation of an arbitration commission is even provided for as another step before approaching the tribunal—which is considered to be the last and final step in resolving disputes. The Soviet Union has also expressed its extreme displeasure at any court that presumably could place both states and private companies on the same legal footing.

Most nations would also be willing to see the Authority empowered with the right of inspection to search for evidence of proper observance of rules. The RSNT reflects this desire in Annex I, Article 13(7): "The Authority shall have the right to inspect all facilities in the Area used in connection with any activities in the Area." Paradoxically, the Soviets have expressed a willingness to have the Authority responsible for inspection and supervision, yet the United States wants to place these responsibilities onto the "Sponsoring Party" (a state) instead.[11]

Just as in revenue-sharing, these issues have suffered from a lack of detailed and extensive consideration thus far. How long it will take to resolve them remains to be seen.

In short, therefore, we see that developing countries generally favor a strong international organization given maximum discretion to determine who will have access to resources and the conditions of such mining, a revenue-sharing plan determined by the Authority on some basis other than simply net profits, and the provision for strong Authority inspection and rule-observing functions. The industrial nations, and particularly the mining nations, are far less sanguine about giving the Authority so much power. The only exception to this characterization is in the settlement of disputes, where the mining nations insist on formulating compulsory arbitration.

However, the question of who exactly will do the mining has dominated all other issues at UNCLOS, and it is the issue to which we next turn.

The Mining System

Ambassador Pardo's conception of the proper institutional structure to act as a "trustee" for the mining of manganese nodules was of a rather standard regulatory model which would monitor activities in the Area and would grant exploration and exploitation rights and licenses to those wishing to exploit resources there. He stated in 1967, "Our long-term objective is the creation of a special agency with adequate powers to administer in the interests of mankind the oceans and the ocean floor beyond national jurisdiction. We envisage such an agency as assuming jurisdiction not as a sovereign but as a trustee for all countries over the oceans and the ocean floor. The agency should be endowed with wide powers to regulate, supervise and control all activities on or under the oceans and the ocean floor".[12] Pardo envisioned, therefore, a strong international regulatory body to function on behalf of mankind in general.

Although willing to examine the issues Pardo raised, the international community was not ready to legitimize his preferred solution—particularly that relating to the proper international structure for mining. Despite ten years of negotiation on this issue, international consensus has yet to form. As was noted previously, the primary issue on the agenda in the late 1960s was the delineation of national from international boundaries. Most nations wished to postpone discussion of the regime's proper institutional structure until the territorial issue was settled. Most sentiment on the subject of institutions, therefore, was carefully qualified by statements regarding its preliminary nature. Løvald states, however, that even in the earliest days of debate developing nations generally supported the creation of a strong supranational organization and industrial nations were definitely less supportive.[13] Table 11, taken from Løvald's analysis, provides a glimpse of this split among the early seabed committee members.

From the beginning there were wide differences of opinion over what would constitute the proper regime. At one end of the spectrum were the Soviet Union and its Eastern European allies, which denied the necessity for any form of international institution. These nations supported the concept of a new international regime to govern deep seabed mining but one without institutionalization. This regime would be no more than a generally accepted set of rules to be observed by the international community. It was, in part, the long-standing ideological objection of the Soviet Union to subordination of the state to international authorities that determined its position. As was pointed out earlier, the mention of "international machinery" in the 1970 Declaration of Principles led to Eastern European abstentions in the U.N. vote. Luard also cites Soviet objection to the formation of a strong international organization on the basis of fears that it would be dominated by the "large-scale capitalist monopolies."[14] Ironically, developing nations have advocated the creation of a strong international organization precisely to prevent "capitalist" monopoly over the seabed.

Since 1970, the Soviet Union and its Eastern bloc allies have modified their position and now regard some form of organization to regulate deep seabed

TABLE 11. Views of Delegates on the Most Appropriate Type of International Regime and Machinery According to Geographic Group

GEOGRAPHIC GROUP	PREFERRED TYPE OF INTERNATIONAL REGIME AND MACHINERY				
	Compre-hensive	*Interme-diate*	*No Regime*	*Unde-cided*	*Total*
African	6	0	0	0	6
Arab	2	0	0	1	3
Asian	4	1	0	0	5
Latin American	2	3	0	0	5
Eastern European	0	0	3	2	5
Western European and Others	2	6	0	0	8
Nongroup members[1]	2	0	0	0	2
Total	18	10	3	3	34
Percent	52.9	29.4	8.8	8.8	100.0

SOURCE: Johan Ludvik Løvald, "In Search of an Ocean Regime: The Negotiations in the General Assembly's Seabed Committee 1968–1970," *International Organization,* 29, 3 (Summer 1975), 691 (© 1975 by the Board of Regents of the University of Wisconsin System). Reprinted by permission.
[1] Malta and Yugoslavia.

mining as inevitable, if not necessary. No nation now objects to the formation of an international organization per se.

The middle of the spectrum, or what Løvald in Table 11 called "intermediate," was in the late 1960s and until today favored primarily by the mining nations. The "intermediate" position goes beyond simply endowing the Authority with a variety of service functions (e.g., notification of mining intentions and plans, the collection and dissemination of data, etc.); instead, it is empowered to "license" qualified mining contractors and authorize and establish some of the conditions or rules of mining, but not engage in the mining activity itself through operational activities. The United States has favored the licensing approach to mining from the earliest days of seabed consideration. Insisting that all qualified mining entities be allowed nondiscriminatory access to the deep seabed—thereby limiting the licensing discretion of the Authority—the United States has favored a "first-come, first-registered" licensing system, which might be termed the "registry approach."

Other mining nations, such as Great Britain, France, and Japan, have favored a licensing approach that would give the Authority more autonomy or discretion in licensing than would the registry approach. Under such a system, nondiscriminatory access to the Area would not be guaranteed to all qualified mining companies because national quotas would be set by the Authority. Hence, quotas could limit the number of U.S. mining organizations allowed to mine.

The early 1970 British and French proposals also envisioned what has been called a three-tier structure system. Under this system, the Authority would license sites to states, which would then, in turn, license or lease areas under their jurisdiction either to private or public mining entities. Hence there would be a complex interaction among international, national, and private interests and activities.

What these various licensing proposals provide is a minimum degree of international control over activities in the Area once licensing has been consummated and a maximum of discretion being left to the actual private or public mining organizations.

At the other end of the spectrum from no international machinery (what Løvald in Table 11 calls "comprehensive") is the operational, or enterprise, system whereby the Authority itself is vested the exclusive right to explore and exploit the resources of the Area. There are also variants of this system, in the most extreme form of which the Authority would raise its own capital and develop its own expertise, completely excluding state and private enterprise from the Area. A less extreme form would be to retain effective and direct control over mining operations but to award service contracts to existing mining enterprises to carry out activities on its behalf. In addition, the Authority could, if it so desired, enter into joint ventures with mining enterprises for specific tasks and exploitation of resources. Regardless of the form, the pervasive influence and control of the Authority would be the prominent characteristic of this system.

The rationale for favoring either the licensing or operational system relates again to one's goals in the Area. If the purpose is to mine commercially as quickly and efficiently as possible, it makes little sense to add another layer of regulatory bureaucracy upon the mining effort—a bureaucracy that no doubt would be large and unwieldy and, like many other international organizations, not held strictly accountable for its actions. Viewed from this perspective it seems senseless to endow the Authority with expertise (not to mention capital) when the task can be carried out by existing firms. If economic equity is an overriding concern, say those favoring the licensing system, the solution is not to impede the functional efficiency of mining but instead to propose an acceptable form of revenue-sharing, whereby those nations in greatest need can reap some of the immediate benefits.[15]

For developing nations, though, such a system (regardless of the extent of revenue-sharing) cannot fulfill the aspirations expressed in Principle VII of the Declaration of Principles:

> The exploration of the Area and the exploitation of its resources shall be carried out for the benefit of mankind as a whole, and taking into particular consideration the interests and needs of the developing countries.

To most developing nations, their perceived interests and needs can only be assured through general international control over the administration of the Area. There is the fear that if decisions and actions are left to the discretion of traditional mining entities, little express attention will be devoted to the concerns of the developing nations.[16] Paul Engo, chairman of UNCLOS Committee I, stated, "The developing countries fear that so long as these companies have guaranteed access and alone possess the necessary finance and technology they would dominate seabed mining in a monopolistic manner. This would deprive the rest of the international community of any significant role in seabed mining. The developing countries envisage the Enterprise as a suitable means for offsetting such a monopolistic situation and for achieving this meaningful role."[17]

The licensing system, then, because it leaves a preponderance of discretion in the hands of mining organizations or states, has been rejected outright by the developing nations. The representative from Kenya has stated that a licensing system would result in "a common heritage of multinational corporations" and not a common heritage of mankind.[18] President Echeverría of Mexico expanded upon this theme even more forcefully at the Caracas session:

> The granting of concessions to States, or worse, to private, probably transnational corporations, for the exploitation of ocean resources, would be the equivalent of permitting the distribution and occupation of vast underwater territories by a few countries, thereby creating a new form of colonialism for the benefit of the technology and financially advanced countries and converting what is supposed to be common patrimony into a profitable business for a privileged few.[18]

It is interesting to trace how national positions have changed over time. Data collected primarily by Friedheim and Durch allow us to make some generalizations. The most striking fact is the extent to which the term "comprehensive regime" has changed. In the late 1960s only two developing nations were thinking of endowing the Authority with exclusive mining rights to the seabed.[20] What was termed a "comprehensive" approach at the time was the combination of a licensing and enterprise system, whereby the Authority would possess its own operating arm but also license state or private mining entities—thereby allowing simultaneous mining by both international and national (or private) firms. This could be called, therefore, a "mixed system."

Industrial nations were almost unanimously opposed to the mixed system in the 1960s, preferring the particular variants of the licensing system described earlier. Little did they know that in moving toward a mixed system some seven to ten years later they would remain in the middle of the spectrum, because of the increasing devotion of developing nations to the far end of the spectrum. What is particularly extraordinary, in light of what has subsequently occurred, is the developing nations' lack of confidence, made evident by them to Løvald in the late 1960s, that they would even be able to negotiate a mixed system. Table 12 shows that although eighteen national delegates (most developing nations) preferred what was then termed a "comprehensive" regime and machinery (the mixed system), only two delegates thought such an outcome was *likely*. There was, then, a feeling among developing nations that although an Authority with

TABLE 12. Views of Delegates Concerning Likely Future International Regime and Machinery According to Preferred Type of Regime and Machinery

PREFERRED TYPE OF REGIME AND MACHINERY	LIKELY FUTURE REGIME AND MACHINERY					
	Comprehensive	*Less Comprehensive*	*Registry System*	*Not Predictable*	*Total* N	%
Comprehensive	2	9	0	7	18	52.9
Intermediate	0	6	1	3	10	29.4
No Regime	0	0	2	1	3	8.8
Undecided	0	1	0	2	3	8.8
Total	2	16	3	13	34	
Percent	5.9	47.1	8.8	38.2		

SOURCE: Johan Ludvik Løvald, "In Search of an Ocean Regime: The Negotiations in the General Assembly's Seabed Committee 1968-1970," *International Organization,* 29, 3 (Summer 1975), 692 (© 1975 by the Board of Regents of the University of Wisconsin System). Reprinted by permission.

an operating Enterprise would be desirable, getting the industrial nations to accede to such a system was improbable.[21]

Consequently, their goal in the early seabed negotiations was limited (in relation to what was to follow during the 1970s), and there was not much confidence that even this limited goal could be achieved. What has happened to change the relevant terms of reference since then will be discussed in Chapter Ten, but it is important to assess now some of the fundamental factors accounting for it.

In the early 1970s there was a flurry of national proposals dealing with the prospective seabed regime and suggestions therein for its institutionalization. Of particular interest was the August 4, 1971, draft proposal submitted by a group of Latin American countries (Chile, Colombia, Ecuador, El Salvador, Guatemala, Guyana, Jamaica, Mexico, Panama, Peru, Trinidad, Tobago, Uruguay, and Venezuela).[22] It was in this proposal that the concept of the Enterprise was first raised and supported (Article 15):

> The Authority shall itself undertake exploration and exploitation activities in the Area; it may, however, avail itself for this purpose of the services of persons, natural or juridical, public or private, national or international, by a system of contracts or by the establishment of joint ventures.

Article 33 went on to discuss the concept in detail:

> The Enterprise is the organ of the Authority empowered to undertake all technical, industrial or commercial activities relating to the exploration of the Area and exploitation of its resources (by itself, or in joint ventures with juridical persons duly sponsored by States).

It should be remembered that the Latin Americans in the late 1960s were not among the more fervent supporters of a comprehensive seabed regime. Their interests were more clearly focused upon territorial extensions and the legitimization of 200-mile zones of national sovereignty. Why is it, then, that the Latin Americans in 1971 should put forward the most comprehensive proposal for international control of the Area? The answer lies again in the nature of the UNCLOS proceedings, and more specifically in the possible trade-offs that exist among issues. Also within this proposal was a section calling for the establishment of a 200-mile EEZ. In short, the proposal called for two crucial elements which ran counter to the 1970 U.S. proposal: (1) extended national jurisdiction from the coast and (2) strong international control over mining in the Area. These two elements were consciously placed within the Latin American proposal to set up a *quid pro quo* arrangement with the United States. In other words, the Latin Americans made it clear that they would be willing to make the Enterprise proposal "expendable" if the United States and other nations would agree to what they really wanted, i.e., the 200-mile EEZ.

What the Latin Americans did not anticipate, however, was the degree of third-world allegiance that would accrue to this bargaining chip. Third-world devotion to the Enterprise as the *sole* institution controlling the exploitation of the Area did not arise overnight. By the end of 1973, Friedheim and Dutch report, still only fifteen nations supported a strong, exclusive Enterprise-based system.[23] The real catalyst for wide acceptance was the 1974 Caracas UNCLOS session, which marked the formation of the G-77 as a special bargaining interest for the third world. During the years 1974-75, again according to Friedheim and Durch, no fewer than sixty-four nations adopted the Enterprise approach, and the center of the spectrum shifted from a licensing approach to a mixed system.

It was at the Caracas session in 1974 that the United States officially acquiesced in the concept of a 200-mile EEZ. What did not take place at Caracas, however, was a similar acquiescence by Latin Americans, and the third world in general, to the preferred seabed regime of the mining nations. Instead, the G-77 adopted the original Latin American Enterprise scheme in its entirety by stating in a document issued at Caracas,

> All contracts, joint ventures or any other such form of association entered into by the Authority relating to the exploration of the Area and the exploitation of its resources and other related activities shall ensure the direct and effective control of the Authority at all times, through appropriate institutional arrangements.[24]

The G-77 position was prominently featured in the SNT that resulted from the 1975 Geneva UNCLOS session. Article 22 stated unequivocally, "Activities in the Area shall be conducted directly by the Authority." The article continues:

> The Authority may, if it considers it appropriate, and within the limits it may determine, carry out activities in the Area or any stage thereof through States Parties to this Convention, or State enterprise, or persons natural or juridical

> which possess the nationality of such States . . . by entering into service contracts, or joint ventures or any other such form of association which ensures direct and effective control at all times over such activities.

Furthermore, Article 22 went on to say that "as early as practicable" the Authority should identify ten economically viable mining sites and enter into joint ventures concerning them.

Many within G-77 considered Article 22 to be an acceptable compromise to full and total Authority control The use of existing mining companies through service contracts or joint ventures is something less than a "pure" Enterprise system, whereby only the Enterprise would be allowed to explore and exploit. Officials from mining nations and corporations, however, were strongly opposed to such a compromise. There certainly was no obligation under Article 22 for the Authority to enter into any form of agreement with existing mining groups—save for the initial ten sites. Spokesmen from developing nations sought to reassure mining representatives that their elimination from the Area was not contemplated, but such reassurances were not sufficient, as witness the testimony of Kennecott's Dubs before Congress:

> The 77 are trying to buy the developed countries off by offering ten sites under a "joint venture" scheme to get things started. After that, they would presumably have acquired funds, technology and management from us so that we could be quickly and quietly removed from further seabed activity in the future.[25]

The dramatic change in national perspectives from 1967 to 1975 can best be summarized by Figure 1 (based on the Friedheim-Durch data).[26] The years 1967-70 saw large international support for pure licensing schemes or some mix of licensing and Authority operation. The years 1971-73 saw strong support for a mixed system, whereas in 1974-75 overwhelming international support (at least measured by the number of nations) favored a highly supranational organization. Why this change occured over time will be explored in Chapter Ten.

As the center of the supranational spectrum moved further away from the desires of mining nations, there was universal recognition that continued adherence to a simple registry or licensing model was no longer feasible. Consequently, during the 1975 Geneva UNCLOS session, both the Soviet Union and the United States made new proposals which went beyond their previous stands. Both proposals were based on dividing the Area into separate international and national jurisdictions.[27]

The Soviet Union, early in the Geneva session, proposed what has been termed a "parallel system," whereby a portion of the Area (percentage not designated in the draft text) would be put under the sole jurisdiction of the Authority to manage and exploit as it alone sees fit. The balance would be reserved for exclusive state use (or states would license private mining entities). Under this plan, private and state mining enterprises would be assured of access to nodules through the portion of the seabed reserved for state control.

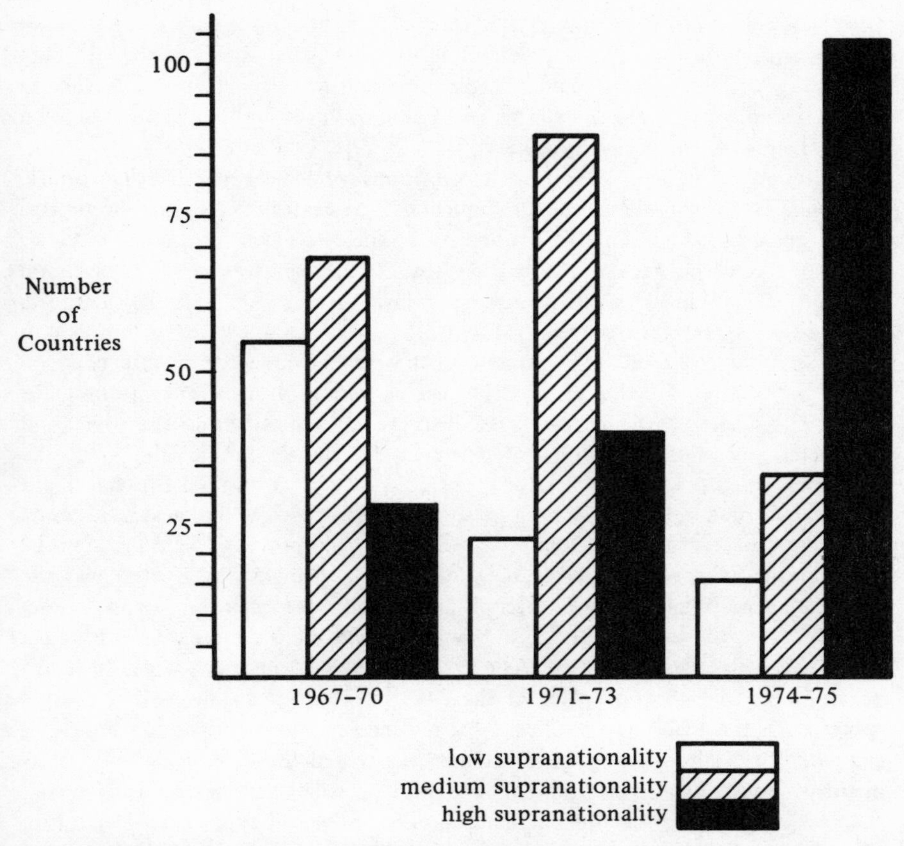

SOURCE: Data from Robert L. Friedheim and William J. Durch, "The International Seabed Resources Agency Negotiations and the New International Economic Order," Paper presented at the 1976 Annual Meeting of the American Political Science Association, Chicago, Ill., Sept. 2–5, 1976.

FIGURE 1. International Support for Supranationality over Time

The United States offered a more detailed and specific version of a regime based on divided jurisdiction. This plan was termed a "banking system," or "twin area system," whereby 50 percent of the Area would be subject to joint ventures between mining entities and the Authority. A maximum of operating discretion would be accorded state or private mining enterprises in this percentage of the Area, and yet basic conditions of mining would be established in the treaty itself. The other 50 percent would be reserved to the Authority for either direct Enterprise exploration and exploitation or joint venture contracts, which would be freely negotiated between the Enterprise and state or private mining corporations. Under this system, the state or private corporation would submit two site choices to the Authority, and the Authority would choose which of the sites to leave to negotiation and which to cede to the mining organization under established conditions. The understanding would be that work on the site, subject to established terms, could commence immediately after the Authority made its choice, whereas the timing of exploitation of the "twin" site would be under the discretion of the Authority.

Although both the U.S. and Soviet proposals were steps toward compromise, they were not favorably received by most G-77 delegates at Geneva, who by this time objected to the vision of a divided Area. The 1970 Declaration of Principles had stated that the Area, not only the resources thereon, was to be the common heritage of mankind. Its prospective split into various jurisdictions, therefore, was viewed as a retreat from that principle (the Algerian delegate later remarked, "Are we here to reserve 50 percent of the seabed area for private corporations?").[28] Moreover, the proposals were viewed as transparent attempts to escape any dealing with the Authority. Strong G-77 opposition to the Soviet and U.S. plans was reflected by the SNT in regard to "who shall mine" the Area.

During the 1975–76 intersessional period, however, a number of what might be termed "moderate" developing nations indicated their willingness to accept a "banking," or "parallel," system, provided other compromises were made in the negotiations. The most notable delegate willing to make this switch was the UNCLOS Chairman of Committee I Paul Engo of Cameroon. Engo, consequently, took the lead in the 1976 New York UNCLOS sessions in pressing for acceptance of a parallel system. As a result, a revised banking system was incorporated into the RSNT in place of the G-77 preferred position. As described in Annex I of the RSNT, an applicant for mining (a state or private corporation) can either (1) choose one large mining area which the Authority will subsequently split, leaving one-half to be exploited solely under the Enterprise's auspices and placing the other half essentially under the applicant's control; or (2) submit two separate sites, one of which the Authority will retain for Enterprise exploitation and the other placed under the control of the applicant.

Inclusion of the above parallel system of mining into the RSNT by no means decided the issue of who will mine. Whether this system represented an acceptable compromise was never fully tested during the 1976 New York sessions. That it still remained the central issue was evident from Chairman Engo's report at the conclusion of the second session: "I wish to emphasize that the one central critical issue which must be solved without delay is that of the system of exploitation. . . . In this connection, I propose that the time between now and the next session of our Conference be used to ascertain the precise limits to which Governments will go on this one single question. No other needs engage us."[29]

Developing nations have generally recognized that the parallel, or banking, system does not compromise U.S. or Soviet interests but instead protects them. The Soviet proposal to split the Area, it is often claimed, relates not to mining per se but to the strong Soviet desire for unhindered military navigation. Reducing Authority jurisdiction to only a portion of the Area would presumably relieve the Soviet concern over access. Such considerations must not be absent from U.S. thinking as well, but there appears a far more straightforward reason; i.e., given the vast magnitude of ore and ore sites, the United States can well afford simply to give half to an international organization. Even under a parallel system, in other words, U.S. companies would still be able to find acceptable mining sites indefinitely. This belief was confirmed in a study prepared for the Ocean Mining Administration of the U.S. Department of the Interior, *Manganese Nodule Resources and Mine Site Availability.* The study projected that somewhere between ten and twenty mining efforts will be in progress through the year 2010. Based upon an examination of exploration data, the study furthermore predicted that there were anywhere from 80 to 185 commercially viable first-generation mine sites—more than enough, therefore, for the probable mining entrants even if the Area were split as proposed in the parallel system. Moreover, there appeared to be ample sites (190 to 460) to satisfy entrants beyond the first-generation stage. The study concluded, "Far into the future, there will be ample mine sites to accommodate late entrants into ocean mining."[30]

The parallel system is the ideal form of compromise for the United States. The mining interests obtain all they want, and concurrently the international community is given exclusive jurisdiction over some portion of the Area. The reason for international skepticism over this scheme, however, should be clear. Having obtained all they want, U.S. mining companies would be in no hurry to operate under Authority, and more specifically, Enterprise auspices. Similarly other mining nations would feel reluctant to work for the Enterprise when there is a better alternative, namely, mining on their own. Hence, it is likely that under a parallel system, the Enterprise, lacking access to know-how and technology, would simply stagnate and wither in an operational sense. The practical effect of a parallel system could be the failure of the Authority to carry out a credible and productive mining operation.

Just as in the RSNT, the parallel system found its way into the ICNT, a product of the 1977 New York UNCLOS session. Unlike the RSNT, however, the latter contained clauses, which from the G-77 point of view, would make the parallel system palatable. In the Annex of the ICNT, dealing with basic conditions of exploration and exploitation, Article 4(c)(ii) states that every applicant for a mining site shall "undertake to negotiate upon the conclusion of the contract, if the Authority shall so request, an agreement making available to the Enterprise under license, the technology used or to be used by the applicant." Article 5(j)(iv) continues, "The Authority may require that the Contractor make available to the Enterprise the same technology to be used in the Contractor's operations on fair and reasonable terms."

These clauses would guarantee that the Enterprise would function effectively, and hence directly address the problem that G-77 delegates have identified with the parallel system. However, because such proposals have been strongly opposed by the U.S. delegation, the ICNT represents virtually no advance in this area over what has come before, and the question of what constitutes the appropriate system of exploitation still eludes the conference.

In conclusion, the development of thought regarding who should explore and exploit the Area has been traced in this chapter. Over the past decade, polarization on this question has characterized the negotiations. A likely basis for compromise could be the adoption of a parallel system, which in effect splits the common heritage of mankind into two. How one produces a parallel system that satisfies both the G-77 and the mining nations, however, has yet to be revealed. Chapter Eleven below will outline a different and, the author believes, a more positive approach to compromise.

Notes

1. David A. Kay, "International Ocean Organizations and Their Regulatory Functions," *Perspectives On Ocean Policy,* NSF, Ocean Policy Project, (Baltimore: Johns Hopkins University Press, 1975), p. 290.
2. Løvald claims that the question of supranationality as expressed in the formation of an international organization was the major issue discussed during the Seabed Committee deliberations from 1968 through 1970. Johan Ludvik Løvald, "Planning the Future of Ocean Space: A Case Study of the United Nations Sea-Bed Committee, 1968-70," Unpublished Ph.D dissertation, Northwestern University, 1973, p. 136.
3. David A. Kay, "Operational Aspects of Managing the Oceans," *The Columbia Journal of World Business* (Spring 1975), p. 32.
4. Eugene Skolnikoff, *The International Imperatives of Technology* (Berkeley, Calif.: Institute of International Studies, 1972).
5. Harold Jacobson, "Technical Developments and Organizational Capabilities," *International Organization* (Autumn 1971), p. 782.
6. Roy Skwang Lee, "Machinery for Seabed Mining," in Francis T. Christy, Jr., et al., eds., *Law of the Sea: Caracas and Beyond* (Cambridge, Mass.: Ballinger, 1975), p. 129.
7. A/Conf. 62/c.1/L.7.
8. Each UNCLOS session since 1975 has produced at its conclusion what are termed "negotiating texts." These are not fully negotiated texts but represent a basis for discussion. No vote has yet to be taken on a negotiating text. The 1975 Geneva session produced the Single Negotiating Text (SNT); the 1976 and 1977 New York sessions produced the Revised Single Negotiating Text (RSNT) and the Informal Composite Negotiating Text (ICNT), respectively.
9. A/Conf. 62/c.1/Wr.5/Add.1.
10. Edward Miles, "An Interpretation of the Geneva Proceedings–Part I," *Ocean Development and International Law* (Winter 1976), p. 11.

11. U.S. Draft, A/Conf. 62/c.1/L.6.
12. *Interim Report on the UN and the Issue of Deep Ocean Resources,* House Committee on Foreign Affairs, 90th Congress, 1st Session, December 7, 1967, p. 285.
13. Johan Ludvik Løvald, "In Search of an Ocean Regime: The Negotiations in the General Assembly's Seabed Committee," *International Organization* (Summer 1975), p. 697.
14. Evan Luard, *Control of the Seabed* (London: Heinemann, 1974), p. 130.
15. All industrial nations—with the exception of the Soviet Union—have agreed in principle to the concept of revenue-sharing. The Soviet Union insists that responsibility for the condition of the poorer nations rests with the former colonial powers and the capitalist countries. Hence it concludes that it should not be expected to provide special compensation from ocean resources to the developing countries. Mark W. Janis and Donald C. F. Daniel, *The USSR: Ocean Uses and Ocean Law,* Occasional Paper # 21, Law of the Sea Institute, May 1974, p. 16.
16. Chairman of Committee I, Paul Bamela Engo, has stated in the *Introduction to the Single Negotiating Text, First Committee* (found in *Status Report on Law of the Sea Conference,* Part 3, hearings before the Subcommittee on Minerals, Materials, and Fuels, 94th Congress, 1st Session, June 4, 1975, p. 1273), "The type of international regime contemplated demands the establishment of an International Authority with wide powers sufficient to perform complex functions. . . . A strong machinery would provide reassurances for about 70% of the world's population in the developing world, who tremble at the crossroads of history with fragile economies and have little faith left in the benevolence of the wealthier nations."
17. A/Conf. 62/L.16, p. 9.
18. Francis Njenga, in *Perspectives on Ocean Policy,* p. 109.
19. Speech by President Echeverría delivered at the Law of the Sea conference, Caracas session, July 26, 1974.
20. Robert L. Friedheim and William J. Durch, "The International Seabed Resources Agency Negotiations and the New International Economic Order," Paper delivered at the 1976 Annual Meeting of the American Political Science Association, Chicago, Ill., Sept. 2–5, 1976.
21. Løvald, "In Search of an Oceam Regime," p. 690.
22. A/AC. 138/49.
23. Friedheim and Durch, "The International Seabed Resources Agency," p. 12.
24. A/Conf. 62/c.1/L.7, p. 1.
25. Marne A. Dubs, *Geneva Session of the Third UN Law of the Sea Conference,* Hearings before the National Ocean Policy Study of the Senate Committee on Commerce, 94th Congress, 1st Session, June 4, 1975, p. 44.
26. Friedheim and Durch, "The International Seabed Resources Agency," pp. 9–12.
27. For a fuller discussion of both Soviet and U.S. proposals, see *Status Report on Law of the Sea Conference,* Hearings before the Senate Subcommittee

on Minerals, Materials, and Fuels, 94th Congress, 1st Session, June 4, 1975.

28. Plenary session, Law of the Sea Conference, May 20, 1976.
29. A/Conf. 62/L.16, p. 7.
30. Ocean Mining Administration, "Manganese Nodule Resources and Mine Site Availability," Study prepared for the Department of the Interior, August 1976, p. 11.

Chapter 7

Decision Making

Although the functions and powers of the prospective ISA have remained issues of considerable contention throughout the negotiations, the basic structural makeup of the organization has not been widely disputed. ISA will consist of both legislative (assembly) and executive (council) organs. Serving the council specifically will be an economic planning commission, a technical commission, and a rules and regulations commission. In addition, a secretariat will carry out the necessary administrative functions, and an Enterprise will be the operational arm. Whereas this framework is generally accepted, disagreement over implementation of the structure still exists and is the subject of this chapter.

There are three major disputes over structure. First, the respective functions of the assembly and the council have been contested, with nations divided over which organ should have primary decision-making powers. Second, there appears to be no consensus regarding which nations should be appointed to the limited-membership executive council. Finally, nations have yet to agree upon appropriate voting systems within both the assembly and the council. How intractable nations are on these three disputes is difficult to assess since UNCLOS sessions have not yet fully focused on the structural dimension.

The preceding chapter was devoted to an examination of varying perspectives on the nature of an appropriate system for mining deep seabed nodules. The issue, in its most basic form, revolved around the question of *who* would control mining activities in the Area (an international body, states, or nongovernment mining organizations). The formation of any organization and the determination of its jurisdictional scope does not resolve the issue of who is in charge. Instead, attention must be focused on the internal composition of the organization, its membership, and the decision-making structure, if one is to determine where control rests.

The question of who (among the international community) will be in charge of the Authority has been an issue that has thus far stayed in the background relative to the issue of supranationality. This fact is perhaps surprising because in a very real sense it is the most crucial political issue to be resolved. The fact that there has not been extensive debate on the subject should not contribute to the impression that it is considered unimportant or that consensus has already been obtained. Nothing could be further from the truth. It remains a latent concern, one suspects, because of the desire of delegates to put first things first. In the late 1960s and early 1970s attention was focused on establishing the jurisdictional boundaries of the organization in a geographical sense. With consensus on this issue, attention shifted to defining the system of mining and the roles of various participants. Since this issue has yet to be resolved, negotiations have not fully moved on to other important and outstanding disputes.

It was stated in the previous chapter that a licensing system has been the preferred approach of mining nations to deep seabed mining. But the framework of an organization does not reveal who is in control. Were developing nations prominently placed within the key decision-making organs of the Authority, they could exercise considerable control even over a licensing system—presuming, of course, that the conditions of access and exploitation were not all mapped out in the treaty itself. A licensing system, therefore, is not of sufficient consequence, in itself, to automatically gain approval from mining nations.

Similarly, if a strong operational Authority were to be created, developed states could still retain effective control over activities within the Area, provided they were accorded a preponderance of power within the decision-making organs of the Authority. In this sense, then, the creation of a strong Authority does not foreclose the possibility of control by industrial nations.[1] The United States, for example, was the prime founder of perhaps the strongest operational international organization yet created, the International Communications Satellite Organization (INTELSAT). Despite the operational nature of INTELSAT, the United States remained in effective control of the organization because, in its early years, it possessed fully 61 percent of the total voting rights. Its voting strength in INTELSAT has now diminished to 33 percent, which still gives it formidable strength when decisions are taken. The point is that control by industrial nations over activities in the Area need not be incompatible with strong international control. As we saw earlier, the Soviets have been against the formation of a strong international organization in part because of its fear that such an organization would be dominated by the industrial, capitalist powers.

Control by developing nations is contingent upon both the creation of a strong international organization and dominance within its decision-making organs. That strong international Authority with control over decision-making by industrial nations would be no victory at all for the developing nations has been perceived by Venezuelan Ambassador Aguilar:

> Were the developing nations to take a purely passive position as beneficiaries, this in my view would condemn them to a situation of perpetual minority or dependence. The rights and interests in the common heritage would for an extended period be administered by the more developed nations, and the developing nations could be relegated to the role of mere marginal spectators in the development process of the Area's resources.[2]

The revolutionary nature of the G-77 goal should be well understood. The Group not only wishes to create an international organization of unprecedented operational authority in a single functional area but also insists that the composition be structured in such a way that those nations with advanced technical capabilities are *not* in a position to dominate. Up to the present, those nations possessing technical capacities and the largest resources have been able to greatly influence, if not control, decisions emanating from limited-member, decision-making organs of functional international organizations. Pelcovits claims that despite pressure to accommodate an expanding number of developing nations into the decision-making bodies of technical and functional international organizations, representation is still largely skewed to those nations possessing the greatest technical skills.[3] G-77 is proposing a complete change from this practice by insisting that those nations possessing capital and expertise be prevented from exercising dominant control. Viewed in this light, it is little wonder that mining nations have so stubbornly resisted their radical designs.

The Caracas session of the Law of the Sea Conference did not deal with the question of the Authority's decision-making bodies (in terms of both composition and voting) but instead deliberated over the system of exploration and exploitation, the enumeration of basic rules or conditions, and the potential economic impact of commercial mining. At the 1975 Geneva session, however, the issue of the Authority's machinery was addressed directly and in some depth. There was general agreement on the creation of a number of organs, such as the traditional assembly, formed as a large, quasi-legislative body; an executive council with fewer participants than the assembly; a secretary-general; and a seabed tribunal. It was also generally agreed that further bodies or commissions could be formed based on the scientific or economic expertise of selected members.

Discussion at UNCLOS sessions has centered on the two most important organs in terms of decision-making, the assembly and the council. For this reason, the rest of this chapter is devoted to an analysis of this dispute. The controversy can best be revealed by breaking it down into two sections: (1) an examination of the policy-making functions or powers of both the council and assembly and an assessment of the relationship between them, and (2) an analysis of the potential composition of the Council and its voting procedures.

Functions and Powers of the Assembly and Council

The very basic question of which Authority organ is to be endowed with primary responsibilities for policy was not resolved at Geneva. It was generally agreed that the assembly should be composed of each nation signatory to the treaty and that each nation should be accorded a single vote. The council, on the other hand, was to be composed of anywhere from twenty-four to forty-eight members. Dispute over which of these two bodies was to be the primary policy arm divided again along North–South lines. The developing nations, stressing the concept of sovereign equality of all states, argued that major Authority decisions needed to be made by the international community as a whole, i.e., the assembly, and not by a small representative body such as the council. The developed nations, however, stressing the efficiency of a smaller body and accustomed to dominating the composition of an executive organ, argued in favor of vesting power primarily in the council.

The SNT, which elaborated the G-77 preference, promoted a strong assembly. Article 26(1) stated,

> The Assembly shall be the supreme policy-making organ of the Authority. It shall have the power to lay down general guidelines and issue directions of a general character as to the policy to be pursued by the Council or other organs of the Authority in any questions or matters within the scope of this Convention. It may also discuss any questions or any matters within the scope of this Convention and make recommendations thereon.

Article 26(2) continued with a listing of explicit assembly powers and functions, some of the most important being the following:

- (i) Election of the members of the Council in accordance with Article 28.
- (iv) Assessment of the contribution of Parties to this Convention as necessary for meeting the administrative budget of the Authority.
- (v) Adoption of the financial regulations of the Authority, including rules on borrowing.
- (vi) Approval of the budget of the Authority on its submission by the Council.
- (vii) Adoption of its rules and procedure.
- (x) Adoption of criteria, rules, regulations and procedures, for the equitable sharing of benefits derived from the Area and its resources, taking into special account the interests and needs of the developing countries, whether coastal or land-locked.

Besides listing the above functions, the SNT left the door open for expanded assembly power by stating in Article 26(3), "The powers and functions of the Authority not specifically entrusted to other organs of the Authority shall be vested in the Assembly."

Each member of the Assembly would have one vote, and all decisions on questions of substance would be made by a two-thirds majority of the members present and voting.

The proposed council's powers and functions, as the executive organ of the Authority, were described in Article 28 of the SNT, which explicitly states, "The Council shall act in a manner consistent with general guidelines and policy directions laid down by the Assembly." The council's main task would be to supervise, coordinate, and implement the provisions of the treaty. To further this task, both an economic planning commission and a technical commission would be part of the council machinery to provide needed expertise. The council would approve contracts for the conduct of activities in the Area, on behalf of the Authority, and would approve and supervise the activities of the Enterprise—the organ which would actually execute the activities of the Authority.

The dominance of the assembly within the proposed SNT structure was strongly opposed by the major mining nations. Their representatives came to New York in 1976 prepared to see major changes in this structure. They succeeded in making the following specific changes which were incorporated within the RSNT:

1. Article 25(6) stated that all decisions on questions of substance would be decided by a two-thirds majority of the members of the assembly (in the SNT, all that was required was simply a two-thirds vote of all those members present and voting).
2. Article 25(7-11) contained various "cooling off," or procedural, mechanisms, preventing dominance by developing nations without time for careful consideration.

Representatives of mining nations were unsuccessful, however, in shaking allegiance to the one-nation, one-vote principle, which was retained in the RSNT.

The powers and functions of the assembly were more carefully spelled out in the RSNT than they were in the SNT. Although the assembly was to retain the same functions as explicitly stated in the SNT, many times they were to be circumscribed by action of the council and the provisions of the treaty themselves. Moreover, the power of the assembly as listed in the SNT, Article 26(3), to decide issues not explicitly entrusted to other Authority organs was qualified specifically. On the other hand, much to the chagrin of representatives from mining nations, the assembly in the RSNT was still referred to as "the Supreme organ of the Authority." In short, the RSNT went a long way toward satisfying mining nations—but not all the way.

At the 1977 UNCLOS session in New York, the pendulum swung back to a certain extent as expressed in the ICNT. Decisions of substance to be decided by the assembly reverted to the SNT standard of two-thirds members present and voting (rather than two-thirds of the total membership). This plan, of course, would further the chances of the developing nations dominating the organ. In addition many of the procedural mechanisms inserted in the RSNT to ensure unanimity were either dropped or qualified significantly. The assembly still retained its status as the supreme organ of the Authority and, in fact, was endowed with new powers to independently select the authority's secretary-general and members to the Seabed Dispute's Chamber of the Law of the Sea Tribunal. The ICNT, therefore, more closely reflects the desires of the G-77 than the RSNT and makes compromise more difficult on this very important issue.

Mining nations have taken a united stand against the preeminence of the assembly as a policy organ. Most industrial nations would prefer to see an assembly having only limited powers or, better yet, only recommendatory functions. Traditionally, large one-nation, one-vote assemblies in international organizations have played a subordinate policy role to limited-membership councils.[4] The Group of 77, now consisting of a total of 106 sovereign states, obviously wants to take advantage of its numbers and use the structure of the assembly to maintain direct control over activities in the Area. The key variable in this issue, then, as in other issues, remains the factor of control.

The Council

Because the mining nations believe that the council should be the predominant organ of the Authority, there has been a good deal of consideration concerning this specific body. All nations agree that the council should be the "executive" decision-making organ (which implies a consensus on the council's functions), but how the council will relate vis-à-vis the assembly and the treaty's provisions specifying the conditions of mining remains subject to negotiation.

Regardless of how the dispute over the council's power will be resolved, there has been intense concern about the structure of the council itself, in which there are essentially two issues: First, since the composition of the council is to be limited, hard choices have to be made about who will be the appropriate members. In other words, an acceptable basis upon which to choose national representatives must be devised, as must, second, a voting system acceptable to the international community.

Executive organs of international bodies have traditionally been made up of a small and select number of nations, and early consideration of council membership (e.g., the 1970 U.S. and the 1971 Tanzanian proposals) reflected this pattern. Subsequent proposals reflected more fully the movement during the 1970s for greater participation of developing nations. Table 13 lists a number of proposals for council membership and voting that have been made over years of consideration, first during the Seabed Committee deliberations and later at UNCLOS. Hence, proposals for a thirty-six-member council are now considered an adequate but minimum number, and even a forty-eight-member council is sometimes suggested. As the number of nations increases, of course, the exclusiveness of the organ diminishes.

Far more controversial than the proper number of nations is the question of the proper basis upon which to choose them. Again, the issue breaks down generally on a North–South axis. Developing nations have always pressed for selection of national delegates on the basis of an "equitable geographical distribution" so that there would be appropriate regional balance. Presumably, such a division would be based on the familiar U.N. regional breakdown—Western Europe (and "Other"), Eastern Europe, Asia, Africa, and Latin America. Disaggregated regional blocs could also be the basis for representation: South America; Caribbean and Central America; the Pacific and the Far East; East Africa and West Africa; Eastern Europe; Southeast Asia and South Asia; North America; Northwest Europe; and the Mediterranean. Regardless of the specific breakdown, any applied equitable geographical distribution would place a preponderence of developing nations on the council, and were G-77 affiliation to prove a binding force, these nations would collectively represent the major power bloc of the Authority.

TABLE 13. International Proposals for Council Membership and Voting Structure

	TANZANIA, 1971[1]	SOVIET UNION, 1971[2]	LATIN AMERICA, 1971[3]	IRAQ, 1971[4]	"PINTO" TEXT, 1975[5]	SNT, 1975[6] RSNT, 1976[7]	ICNT, 1977[8]
Membership	18 Chosen on basis of equitable geographic distribution	30 Based on equitable geographical distribution	35 Based on equitable geographical distribution	30 1. 15 developing states 2. 5 highly industrial states 3. 3 other developed states 4. 2 GDSs[9] 5. 5 international agencies whose tasks extend to some form of marine management	36 1. 18 based on equitable geographical distribution 2. 18 on basis of special interests a. 9 from industrial nations making largest contribution to mining b. 9 from developing nations	36 1. 24 based on equitable geographical distribution 2. 12 on basis of special interests a. 6 developing nations b. 6 industrial nations	36 1. 18 based on equitable geographical distribution 2. 18 on basis of special interests a. 4 nations making the largest investments in nodule mining b. 4 major mineral-importing nations c. 4 major land-based mineral exporters d. 6 developing nations
Voting Structure on Matters of Substance (One-Nation, One-Vote)	2/3 majority of all members	Decisions made by agreement, requiring universal consensus	2/3 majority of members present and voting	2/3 majority of members present and voting	3/4 majority of members present and voting	2/3 + 1 of those present and voting	3/4 majority of members present and voting

[1] A/AC. 138/33.
[2] A/AC. 138/43.
[3] A/AC. 138/49.
[4] A/AC. 138/SC.I/SR.5–31.
[5] Hearings before the U.S. Senate, Subcommittee on Minerals, Materials, and Fuels, *Status Report on Law of the Sea Conference*, June 4, 1975, pp. 1251–68.
[6] A/Conf. 62/WP.8/Pt. I.
[7] A/Conf. 62/WP.8/Rev.1/Pt.I.
[8] A/Conf. 62/WP.10.
[9] Georgraphically disadvantaged states.

The North has argued, however, for special interest representation, claiming that the council should primarily reflect the interests of those nations most deeply concerned with or involved in deep seabed mining. With this criterion, the interests of mining nations would be better protected than with equitable geographical distribution. It is now generally accepted that the council should ultimately contain a mix of special interest and regional representation. Unresolved as yet is the exact proportion. The SNT and the RSNT split the council on a two-thirds (regional) and one-third (special interest) basis. The ICNT, however, had a fifty–fifty split. What interests should gain representation is another difficult matter. The following categories have all been suggested at one time or another as deserving special representation: advanced technical expertise in deep seabed mining, advanced work in offshore mining in general, advanced work in marine technology generally, either high or low per capita gross national product (GNP), geographically disadvantaged states (and- or shelf-locked states), large populations, highly industrial societies, socialist governments, large mineral-producing sectors, large mineral-consuming sectors, large mineral-exporting sectors. The problem is trying to produce an equitable and acceptable balance from this mixture of interests.[5]

The SNT and RSNT, as seen in Table 13, provide for a rather complex system of compromise. Out of a total of thirty-six council members, there would be twelve reserved seats. Of these twelve, six would be set aside for countries with substantial investment in ocean mining (or with advanced mining technology) or for the major importers of the minerals produced from the Area (one of the six seats would be required to go to a socialist Eastern European country). The remaining six reserved seats would go to developing nations, one from each of the following categories: land-based mineral exporters, mineral importers, land-locked nations, shelf-locked nations, states with large populations, and least developed nations. The remaining twenty-four seats would be apportioned on a geographical basis—terms lasting for a period of four years—using the following blocs: Africa, Asia, Eastern Europe (socialist), Latin America, and Western Europe and Other.

This composition of the council appears unacceptable to those industrial nations possessing advanced ocean mining technology. In their view, the proposed council would provide inadequate representation for those nations having substantial investment and interest in actual mining operations.

The ICNT comes closer to meeting the desires of mining nations as it reduces the number to be chosen by equitable geographical distribution (from twenty-four to eighteen). This plan would allow for greater representation from industrial interests.

However, who will be on the council does not necessarily answer the question of who will control it. Left unsaid thus far is the importance of the voting system in determining who will wield power. In other words, how votes will be taken on substantive matters must also be subject to serious consideration. Mining nations, for example, might not object to minimum national representation on the council if they could be assured that their interests would be protected through an appropriate voting system. To determine the power base of the organization one must know both national representation and the decision-making process.

The basis for a council voting system has throughout the long years of negotiation been the one-nation, one-vote standard. In a sense, this fact is not surprising since it is the standard associated with most U.N. bodies. In another sense, it is unusual because of the vastly varying capabilities among nations to mine. Despite this incongruity, the one-nation, one-vote principle remains one of the most firmly entrenched demands of the G-77.

In 1967, Maltese Ambassador Pardo expressed another opinion: "I would only observe that it is hardly likely that those countries that have already developed a technical capacity to exploit the ocean floor, would agree to an international regime if it were administered by a body of small countries such as mine, having the same voting power as the U.S. or the Soviet Union."[6] Other observers have also noted that it would be politically inconceivable to expect the major mining powers to accept equal voting rights with the smaller, less involved nations.[7] Mining nations have indeed put forward a different basis for voting, as seen in the various U.S. proposals outlined in Table 14. However, the G-77 insistence upon the "sovereign equality of all nations," translated practically into a one-nation, one-vote standard, has consistently appeared in the negotiating texts emanating from the 1975-77 UNCLOS sessions.

Similarly, the requirement that substantive decisions be reached by a two-thirds majority of those council members present and voting is another common U.N. standard having the full support of the G-77. Friedheim and Durch claim that their data, obtained for the years 1971-73, indicate very strong support for the traditional two-thirds present and voting majority.[8] However, other standards have been proposed. Perhaps the most unusual and certainly the most stringent one would be the Soviet "consensus" system proposed in 1971. This proposal, which would require universal acceptance of the question at issue by all council members, has not been heard since the early 1970s.

A somewhat stiffer requirement than the two-thirds majority would be that for a three-fourths present and voting majority. Such a standard has been found in both the "Pinto text" (an informal text circulated during the 1975 Geneva session, forged by the delegate from Sri Lanka, Christopher Pinto) and the 1977 ICNT. The difference between a two-thirds majority and a three-fourths majority can be significant. Given the representation within the SNT (twenty-four nations chosen by equitable geographical distribution and twelve by special interest), it is likely that the council would consist of at most thirteen industrial nations. Under a two-thirds majority rule, industrial nations, voting as a bloc, could stop any proposal detrimental to their perceived interests. Representatives from the United States and other industrial nations, however, have indicated that this thin margin of protection is unacceptable—particularly since these nations are accustomed, on the basis of their power, expertise, and investment, to

TABLE 14. U.S. Proposals for Council Membership and Voting Structure

	1970[1]	1975–76[2]	1976[3]
Membership	24 1. 6 of the most industrial advanced nations 2. 18 by equitable geographical distribution (at least 12 of which would be developing nations)	36 1. 6 industrial nations making the greatest contribution to mining 2. 6 developing nations 3. 6 land-based mineral producers 4. 6 major consuming nations 5. 12 by equitable geographical distribution	36 1. 24 on basis of special interests a. 6 industrial nations making the greatest contribution to mining b. 6 developing nations c. 6 land-based mineral producers d. 6 major mineral consumers 2. 12 by equitable geographical distribution
Voting Structure on Matters of Substance (One-Nation, One-Vote)	A majority of the council members, as well as a majority of members in each of the two categories above	3/4 vote of those present and voting; simple majority required in at least 3 of the above categories	3/4 vote of council members, provided this represents more than one-half of the total value of production and consumption of the minerals derived from the area.

[1] A/AC. 138/25.
[2] *Proposed Amendments to the Committee I Single Negotiating Test,* Dec. 1975.
[3] From the U.S. informal text, Sept. 7, 1976.

playing a far greater role in the policy formation of international organizations. Given the representation designated in the Pinto Text (eighteen chosen by equitable geographical distribution and eighteen by special interests), it is likely that the council could consist of as many as fifteen industrial nations.[9] With a three-fourths majority, industrial nations would almost certainly have a blocking vote on substantive decisions. The ICNT resembles the Pinto text in both representation and voting system, and it is quite possible that this formulation will find its way into the final treaty.

The United States has consistently pushed for a voting system that would give industrial nations what would, in effect, be a veto over council decisions (see Table 14). The 1970 U.S. proposal required approval of substantive decisions by a simple majority; but a majority in each of the two separate categories—special interests of industrial nations and regional representation—would also be required (a concurrent majority). Consequently, all eighteen members elected by equitable geographical distribution could vote for a measure, but if it did not gain the approval of more than three industrial nations, it would not pass. Given the strict adherence of developing nations to the one-nation, one-vote principle, it was not surprising that this U.S. proposal drew little support.

The United States was not discouraged from proposing similar schemes, however. The 1975-76 proposal would split membership into five separate groups, with passage of issues required by a majority in at least three of the categories and by a three-fourths vote of those present and voting overall. The 1976 proposal, which was circulated during the summer New York session, had the same representation as the previous one but with a slightly different twist for voting. A three-fourths vote of *all* council members would be required for passage (much more stringent than simply a three-fourths vote of members present and voting) if this vote included states having more than one-half the total value of consumption and production of the minerals to be derived from the Area. This latter clause was inserted to ensure that major mineral-producing and -consuming nations had voting power "commensurate" (a buzz word of the mid-1970s) with their interests in the minerals.

Although each of the U.S. proposals shown in Table 14 have slightly different voting systems, they all had the same purpose, i.e., to provide the United States, in association with a few allies, the power to block resolutions favored by the G-77. Each of the formulations would have accomplished this task, as would probably the ICNT. Nevertheless, the U.S. formulations provide firmer assurances against the "automatic majorities" of the G-77.

To many members of the U.S. delegation it has been a source of irritation that the United States, as the most significant marine mining participant, should be reduced to simply a blocking role within the council voting system. They have instead pushed for the power to produce "positive," or "affirmative," decisions. The only way to do so would be to replace the one-nation, one-vote principle with a system of weighted voting, for which there are precedents in existing international organizations. In the World Bank and the International Monetary Fund (IMF), for example, each nation's voting share is based on its level of investment in the two organizations, and consequently the richer nations control their functioning. An even better analogy to the proposed Authority is INTELSAT, which deals with communications satellites and their transmission. Voting is based on investment levels in the organization itself, and thus the United States and Western European nations play a dominant role.

Still other precedents exist in various commodity agreements, e.g., the International Coffee Agreement, the International Tin Agreement, and the International Wheat Agreement, which all have been established by the major producers and consumers of the respective commodities. Voting rights are accorded to nations on the basis of their importance in both the exporting and importing of the commodities. In the International Coffee Agreement, for example, the

United States holds close to 40 percent of the consumer vote because of the large share of coffee it imports.

Weighted voting, therefore, would give the United States far greater protection of its interests than any other system yet described, and in some cases would allow the United States and its allies to dominate the voting system of the organization. Thus, the Treasury Department in 1976 made a strong effort to have such a system adopted within the U.S. delegation. They chose as a model the commodity arrangements, where votes are accorded on the basis of the quantity produced, consumed, or traded. The Treasury Department calculated that were voting shares to be accorded by total consumption and production of the four nodule minerals of commercial value, the United States would receive 16.1 percent of production votes and nearly 25 percent of all consumption votes—or if combined, approximately 20 percent of all votes.[10] Were a three-fourths vote required for decision-making, the United States would only need the weighted votes of the Soviet Union, Japan, or Canada to block any measure. Moreover, if votes were weighted according to production and consumption and then combined, the United States, Western European nations, Japan, and the Soviet Union could together amass 70 percent of the council votes—thereby giving the mining nations a strong chance for affirmative action. The chances would be increased, of course, if only a two-thirds majority of those present and voting were required.

Although this proposal gained serious consideration within the U.S. delegation, it never was advanced to the status of an informal draft. There was, in fact, little chance for international acceptance. It had become apparent by 1976 that the international community was unwilling to accept the concurrent majority voting system which the United States had advocated up until then. What made the Treasury Department think that the international community would accept a weighted voting system—a system far less acceptable than a concurrent majority system—has never been explained. Although no one would deny that the weighted voting system would give the United States more power within the council than any other system, it was never politically feasible. U.S. representatives, recognizing this fact of life, therefore never brought it forward for international consideration.

Conclusion

There is, perhaps, no more evident indication of the changing nature of international relations today than the insistence of developing nations in playing a crucial role in international organizations. Had UNCLOS negotiations taken place in the mid-1960s, there would have been little question about who would be in control of the decision-making organs of the Authority. In 1967, as Pardo stated, it was politically unthinkable that the United States and states such as Malta, Zambia, and Tonga would be accorded equal voting rights in a functional international organization. Ten years later it became not only thinkable but a probable outcome of seabed negotiations. Should the Authority generate enough revenue from seabed mining to support itself, the one-nation, one-vote standard should not prove to be a problem. If, however, the Authority requires extensive state funding for its support, that standard could prove costly, in the sense that such nations as the United States, France, and the Soviet Union will feel little compulsion to contribute more than a nominal amount. Control by developing nations of international organizations carries its own risks for the developing nations themselves.

Notes

1. Despite this fact, the chief U.S. negotiator in Committee I suggested in congressional testimony the unacceptability of such an exchange: "We cannot, in my view, afford to trade a satisfactory resource Authority for an unsatisfactory system of exploitation." *Status Report on Law of the Sea Conference,* Hearings before the Senate Subcommittee on Minerals, Materials, and Fuels, 94th Congress, 1st Session, June 4, 1975, p. 1190.
2. Andres Aguilar, "How Will the Future Deep Seabed Regime be Organized?", John K. Gamble and Giulio Pontecorvo, eds., *Law of the Sea: The Emerging Regime of the Oceans* (Cambridge, Mass.: Ballinger, 1974), pp. 47–48.
3. Nathan A. Pelcovits, "Decision-Making Mechanisms for the International Seabed Resource Authority: The Impact on U.N. Interests," Report prepared for the U.S. Department of State, April 30, 1976, p. 35.
4. Eugene B. Skolnikoff, *The International Imperatives of Technology,* (Berkeley, Calif.: Institute of International Studies, 1972), p. 137.
5. Clearly, not all interests can be fully satisfied. Czechoslovakia, acting on behalf of the Group of Land-locked and Geographically Disadvantaged States at Geneva submitted a proposal declaring, "at least 2/5 of the members of the Council shall be representatives of land-locked and geographically disadvantaged States." In fact, these nations will probably be lucky to capture one-fifth of the total council seats.
6. *Interim Report on the UN and the Issue of Deep Ocean Resources,* House Committee on Foreign Affairs, 90th Congress, 1st session, December 7, 1967, p. 285.
7. Evan Luard, *Control of the Seabed* (London: Heinemann, 1974), p. 236.
8. Robert L. Friedheim and William J. Durch, "The International Seabed Resources Agency Negotiations and the New International Economic Order," Paper delivered at the American Political Science Association, Chicago, Ill., September 2–5, 1976, p. 15.
9. Jonathan I. Charney, "The International Regime for the Deep Seabed: Past Conflicts and Proposals for Progress," *Harvard International Law Journal,* 17, 1 (Winter 1976), 15.

10. "Analysis of Treasury's Proposal for Weighted Voting in the ISRA Council and Implications of Some Proposed Variations," Unpublished paper, U.S. Treasury Department, April 7, 1976, p. 1.

Chapter 8

Basic Mining Conditions

As seen in previous chapters, the profitability of manganese nodule mining will be affected by such factors as the reliability of technology and the grade of nodules collected. At least equally important will be the rules and conditions under which mining will be promulgated—rules and conditions either detailed in the treaty itself or established by the ISA. The magnitude of the nodule mining effort, therefore, may very much be determined by the investment climate produced. This chapter discusses the three major disputes over actual mining conditions: (1) whether qualified mining organizations will have guaranteed access to nodule fields on a nondiscriminatory basis or will have to operate within national quotas, (2) price and production controls (which expands upon the treatment of this issue in Chapter Three), and (3) the role of ISA's Enterprise and its relationship to existing mining firms.

Consideration over the past decade has also been given to the expansion of ISA's functional tasks beyond its acknowledged mining mandate. This chapter examines developments in UNCLOS concerning an ISA role in marine science beyond areas of national jurisdiction. Also discussed is the possibility of ISA expanding its jurisdictional scope to the water column above the seabed.

Satisfactory conditions must be forthcoming from UNCLOS if existing mining companies are to place both their capital and technical expertise at the service of the Authority. Although mining companies would prefer not to function under the aegis of a strong supranational organization, there is no reason to believe they would not, provided that the profits appear favorable. The establishment of a supranational organization to control mining does not necessarily mean that mining organizations could not enjoy traditional levels of profit or return on investment.

As stated previously, multinational land-mining enterprises have demonstrated considerable flexibility in adjusting to changing control. Where once private control over mining sites was the norm, it is now rapidly becoming the exception. Increasing national control over the initial stages of mineral production has meant, in practice, increasing use of joint (public and private) ventures, service contracts (excluding the equity investment of the mining firm), and international consortia.[1] Mining firms have taken to these new arrangements with varying degrees of enthusiasm. Nevertheless, they have for the most part adapted and are still able to gain acceptable profits.[2]

Mining enterprises would not find themselves in a totally novel situation if they were to operate under the control of a strong international Authority. The basic question remains, however, whether traditional mining enterprises can be assured of a stable investment climate and whether they can receive a reasonable return on their investment.

The importance of framing acceptable basic conditions for both mining companies and the international community cannot be minimized. It would be a hollow victory for the developing nations if a strong Authority were formed, but which then promulgated conditions providing inadequate incentives (or disincentives) to attract mining enterprises. There are substantial technical and economic risks which first generation mining operators have accepted. Compounding them with substantial political risks would place the conduct of these activities in extreme jeopardy. It would take the Authority an extended period of time to build up an indigenous mining capacity without the assistance of the established mining groups. It is also unlikely that several industrial states would prevent their mining interests from mining independently if they were unable to strike a bargain with the Authority. Industrial nations might, for example, form a regime of their own for deep seabed mining or initiate unilateral activities and regulation. At the least, failure of the Authority to enlist mining companies would mean that they would search for commercial nodule sites within the economic zones of various islands or nations and then establish mining conditions with national representatives. Borgese claims that as much as 20 percent of all commercial nodules may fall within national jurisdictions, though most estimates place the percentage at closer to 10 percent.[3] It is reported that rich nodule fields may be found within the jurisdiction of the French Polynesian Islands, Tonga Platform, and Western Samoa.[4] It would be surprising if prospective nodule miners have not already investigated thoroughly the prospects for mining within national jurisdictions, but for understandable reasons, they are not talking about it.

In sum, then, mining companies have more possible options than simply accepting Authority rules or conditions. For this reason it is essential that the Authority and existing mining enterprises come to mutually satisfactory terms if we are to see the principle of the common heritage of mankind expressed practically.

Considerable attention to formulating basic conditions of exploration and exploitation was evident during the latter stages of the Caracas session and throughout the Geneva session. During the former, a First Committee Working Group was formed to focus directly on producing basic conditions which could serve as an annex to the convention. The working group produced a document comparing the conditions put forth in various national or multinational proposals submitted for Committee I consideration.[5] It compared the various proposals with regard to thirty-six separate conditions (which are beyond the scope of this chapter to examine). There is no serious disagreement among nations on several conditions. The most serious differences were outlined by the chairman of the Working committee in a speech before Committee I on March 19, 1975, in Geneva.[6] These differences can be grouped into three broad categories of issues: (1) the process by which mining enterprises gain access to Area resources, (2) the dispute over the application of price and/or production controls, and (3) the appropriate scope of resource activities by the Authority.

Access to Area Resources

The question of who is to mine the resources of the Area, and at what pace, will be determined by the system or rules granting access. The question of access in general has already been discussed as it pertains to the relationship between the Enterprise and the mining companies. In addition to a strong North-South disagreement regarding this relationship—and unlike any issue discussed thus far—there is also a significant split among mining nations themselves over access. More specifically, the United States is virtually isolated in the world community in its insistence upon guaranteed access to the Area for all technically qualified enterprises on a nondiscriminatory basis. Our representatives have repeatedly made it clear in all forums available that such access must be an integral condition of mining if the United States is to sign a seabed convention.[7]

Outside of the United States there is universal sentiment for placing a limit on the number of contracts that can be granted to mining organizations based in any one state—which is usually expressed in terms of a quota. The purpose of placing this barrier to unlimited access appears clear. There is considerable fear that unrestricted access to the Area would place first-generation ocean miners, predominantly U.S. mining interests, at a competitive advantage. The call for equal access to the resources of the Area, therefore, is an attempt to limit the advantages that would accrue to American-based companies because of their technical superiority. U.S. efforts to persuade foreign officials that there are sufficient mining sites for all, even with unrestricted mining, have not been notably successful.

One of the earliest indications that the U.S. desire for free access would be opposed by other than third-world nations was the Western European draft proposal presented at Caracas, which stated that an applicant for prospecting or exploitation could not hold more than six contracts at any one time.[8] The concept of national quotas, however, was introduced in the SNT, Annex I, 8(f) and (g):

> The total number of contracts for evaluation and exploitation entered into by the Authority with a single State Party or with natural and juridical persons under the sponsorship of a single State Party shall not exceed . . . percent of the total area open under paragraph 3, and shall be equal for all States Parties.

> Within the limits specified in sub-paragraph (f) the Council may every year determine the number of contracts to be entered into by the Authority with a single State Party or with natural and juridical persons under the sponsorship of a single State Party in order to give effect to the provisions of articles. . . .

In short, the SNT left it up to the council to determine the pace of granting contracts and that there would be an undetermined limit on the percentage of the total Area that states could mine.

The RSNT contains no equivalent article. Not wishing to bring this contentious issue into the 1976 negotiations, the RSNT, Annex I, 8(e) simply states "The issue of a quota or antimonopoly provision remains to be discussed in the Committee in the future." The issue was the subject of considerable negotiation during the 1977 New York Conference, but it was not ultimately resolved. Still, sentiment appears strongly in favor of some form of quota; the ICNT, 5(1), states, "While the inclusion of a quota or anti-monopoly provision appears acceptable in principle, its detailed formulation has yet to be fully negotiated."

How divisive this issue remains among the mining countries is difficult to assess since it is largely overwhelmed at this time by the serious North–South split. It has, however, seriously impeded the formulation of a common proposal from industrial nations to counter the united bloc of G-77. As will be pointed out in Chapter Nine, the lack of unity among the industrial nations has contributed to the caution with which the U.S. administration has proceeded.

Price and Production Controls

The issue of whether the Authority should have the power to enact price and production controls over mining in the Area has been one of the most debated in Committee I proceedings. Again, there is a strong North–South divergence of opinion on this matter. Industrial nations prefer to grant the Authority no such powers, instead providing for a system of "compensation" to affected land-based developing-nation mineral producers. Developing nations maintain, however, that prevention of serious adverse effects on mining nations is the route to take, which would require the application of price and production controls.

Friedheim and Durch report overwhelming national sentiment for some form of controls through the years 1971–75.[9] The SNT allowed for their imposition (Article 9, 1(b)), stating that development of the Area should avoid adversely affecting the mineral export earnings of developing nations. Later, the SNT empowered the economic planning commission to make recommendations to the council regarding the desirable pace of mining and to examine "appropriate programs or measures, including integrated commodity arrangements, to avoid or minimize adverse effects" on developing nations.[10]

It was in the 1976 New York sessions, however, that this issue came to the forefront. On April 8, 1976, U.S. Secretary of State Henry Kissinger made a major speech on the law of the sea, in which he put forward what many called a major concession on the issue of price and production control.

> The United States is prepared to accept a temporary limitation, for a period fixed in the Treaty on production of the seabed minerals tied to the projected growth in the world nickel market, currently estimated to be about 6% a year. This would in effect limit production of other minerals contained in deep seabed nodules, including copper.[11]

He furthermore stated that the United States would recognize the right of the Authority to participate in any international agreements on seabed-produced commodities. For the first time, therefore, and much to the chagrin of U.S. mining entities the government showed flexibility on this issue. The initiative was widely applauded at New York, and negotiations began immediately on how to hammer out the specifics of the Kissinger proposal. The results can be seen in the RSNT. Article 9(4)(i) cites the right of the Authority to participate in commodity arrangements in accordance with the amount of production for which it is directly responsible. Article 9(4) (ii) calls for an interim period of twenty years (with the option of extending it for another five years) upon the commencement of commercial production, during which total production in the Area would not exceed the projected cumulative growth of the nickel market. This projection, based on the average annual rate of increase in world demand during the twenty-year period before commencement of commercial operations, would be at least 6 percent per annum.

U.S. mining officials, in congressional testimony, expressed their displeasure with the concession the administration was making. All indications were, though, that they could in fact live with the RSNT provisions, and that they were essentially basing their objections upon the precedent of encroachment upon market forces. However, they were not fully prepared for what came out of the 1977 ICNT. In it, Article 150 went far beyond the rather straightforward production restrictions proposed in the RSNT. It concluded that there would be, first, an interim period of seven years in which total production of minerals from nodules would not exceed 60 percent of the cumulative growth of the world nickel demand. This standard would continue to be in effect until a commodity agreement were achieved between mineral producers and consumers ensuring a stable market and a fair price as well. In other words, there would be no movement toward a free-market system following the interim period as envisioned in the RSNT. In addition, reference to a minimum 6 percent cumulative growth figure was eliminated, and a complicated system for determining a projected cumulative figure was put in its stead.

The ICNT, therefore, clearly went beyond what the U.S. government was willing to offer as a concession and as such was unacceptable to both miners and government officials. As a result, UNCLOS is hardly any further ahead now on this issue than it was some seven years ago. The only difference is that then there was disagreement over principle and today there is disagreement over specifics. Judging from the ICNT, there is a considerable distance to be covered if satisfactory resolution is to be achieved.

Scope of Authority Over Resource Activities

Whether the Authority should conduct resource-related activities besides direct mining of the Area has been an issue of some contention. There has been no dispute over *whether* the Authority would have the right to conduct scientific research in the Area, for example, but over how large a role the Authority should play. The SNT called for the Authority to be the "center for harmonizing and coordinating scientific research." As such, it would have full control over all research in the Area, requiring existing oceanographic institutions to submit their plans to the Authority for approval. Moreover, in the SNT, publication and dissemination of research results would have to be through the exclusive auspices of the Authority.[12]

Oceanographers and marine scientists—the overwhelming majority located in the industrial nations—have traditionally enjoyed freedom of scientific research beyond narrow territorial limits. There is now some question whether they can retain this freedom in the face of pressures (emanating from Committee III of UNCLOS) for increased regulation in the EEZ as well as in the Area.[13] Authority control of all research in the Area, therefore, raised the spector of a completely new regime for scientific research throughout the oceans, something that was clearly objectionable to those nations with large and active oceanographic institutions.

In light of strong objections to SNT articles on scientific research, the 1976 New York session witnessed some significant changes. The RSNT made no mention of the Authority being a "center" for scientific research. In its place was the more limited phrase, "The Authority shall promote and encourage the conduct of scientific research in the Area."[14] In addition, dissemination of results was no longer required to be the sole province of the Authority.[15] The ICNT essentially retained the wording and spirit of the RSNT. It appears, therefore, that oceanographers need no longer fear the dominating presence of the Authority in their work.

The foregoing three issues related to basic conditions for exploration and exploitation have been subject to debate since the early 1970s. A separate issue, however, has become prominent only during the latest UNCLOS sessions and now deserves clarification—the establishment of the Enterprise as the operating arm of the Authority.

The Enterprise

The specific actions leading to favorable support for the concept of an Enterprise have been described earlier. Proposed initially by the Latin Americans in

1971 as an expendable bargaining chip, the Enterprise gained the strong support of the G-77 until it became a *sine qua non* of any ultimate settlement of the seabed issue. The United States formally acquiesced in the establishment of an Enterprise in the 1975 and 1976 addresses of Secretary of State Kissinger on the law of the sea.[16] However, the United States and other mining nations made clear that their support of the Enterprise was contingent upon a parallel system of mining, allowing relatively free access to seabed minerals for their own mining entities. It was well understood that the Enterprise was to be a concession of the mining nations in exchange for an acceptable (to existing mining entities) mining system.

Specific consideration of the makeup, functioning, and character of the Enterprise did not begin until 1976. It was not until late into the first 1976 UNCLOS session that what was ultimately to become Annex II of the RSNT on the Enterprise was circulated for consideration. Annex II provided for the establishment of a governing board that would be responsible for the conduct of Enterprise operations. This board would contain members elected on the same basis as the council (i.e., twenty-four would be elected by equitable geographical representation and twelve by the principle of special interests). Similar to the council, a one-member one-vote rule would prevail, but it would take only a simple majority of those present and voting to pass measures. In order to establish the necessary financial footing, the Enterprise would have the power to borrow funds.

During the session, considerable discussion revolved around the financing of the Enterprise. The mining nations generally supported borrowing in capital markets, as well as a certain portion of revenue-sharing funds earmarked especially for the Enterprise. Some developing nations, however, adamantly fought for the right to assess states certain fees in its support.[17] The issue of financing was again taken up at the New York 1976 summer UNCLOS session. There, Secretary of State Kissinger pledged both technical and financial assistance, which would enable the Enterprise to begin operations upon commencement of Authority management. Again, the pledge of assistance to the Enterprise was clearly linked to the requirement for a parallel mining system satisfactory to the United States. Less clear were the specific details Kissinger had in mind when he made the proposal.

If the U.S. government was willing to help in the Enterprise only in exchange for a satisfactory system of mining, many developing nations were willing to acquiesce in such a system only if a competent and strong Enterprise were created. Rhetorically, the goals of the nations were alike or at least not incompatible, but in practice the visions of a functioning Enterprise were strongly dissimilar. This divergence can be seen in the ICNT, which evolved from the 1977 UNCLOS session. In it is a parallel system of mining, much in line with the wishes of the United States and other mining nations. However, the annex of the ICNT dealing with the Enterprise has been altered considerably from the RSNT and constitutes a high price for the mining nations to pay. First, stiff requirements on the transfer of technology would be placed on mining entities.

In the annex dealing with basic conditions of exploration and exploitation, two sections are particularly relevant. Article 4(c)(ii) says that every applicant for a contract area shall "Undertake to negotiate upon the conclusion of the contract, if the Authority shall so request, an agreement making available to the Enterprise under license, the technology used or to be used by the applicant, in carrying out activities in the Area on fair and reasonable terms." Article 5(j)(iv) is even more explicit: "The Authority may require that the contractor make available to the Enterprise *the same technology* to be used in the contractor's operations on fair and reasonable terms." (Author's emphasis.) Given such explicit references it is difficult to see how mining entities could escape the wholesale turnover of "proprietary" information and technology to the Enterprise.

Financial obligations were also spelled out in the Annex dealing with the Enterprise. Article 10 lists the various ways the Enterprise can finance its operations, including borrowing, obtaining Authority funds specifically so earmarked, and obtaining voluntary contributions. This article, however, leaves the obligations of states open-ended, as it claims in 10(a)(vi) that raising sufficient capital could include "other funds made available to the Enterprise including charges to enable it to carry out its functions and to commence operations as soon as possible." No system of dues is explicitly provided for until 10(c)(iii), which states that in the event that Enterprise costs on its first site cannot be raised through the previously listed means, "States Parties shall guarantee debts incurred by the Enterprise for the financing of such costs." The amount states would be liable for would be in proportion to the U.N. scale of assessments. It has been reported that during the 1977 session, the United States offered to finance fully 20 percent of the Enterprise's first mining site—which could amount to a U.S. bill of approximately $175 million. Although the United States was apparently willing to make this large contribution to the Enterprise's first endeavor, it was not willing to submit to the open-ended obligations in the ICNT.

Finally, the governing board in the ICNT has been changed from the RSNT to only fifteen members elected by the assembly. All, not some, are to be chosen by the principle of equitable geographical representation.

In sum, whereas the fundamental bargain on an acceptable mining system appears to have been struck, there remains a serious and wide split over the establishment of the Enterprise. However, the dispute has not been eliminated by the acceptance of the parallel system but simply shifted to focus on the Enterprise.

Scope of the Regime

The United States (and the Soviet Union) has from the beginning of seabed negotiations attempted to narrowly restrict the functional jurisdiction, or scope, of the proposed Authority. It should be pointed out that the deep oceans, beyond projected limits of national jurisdiction, are more than simply a reservoir for manganese nodules. Numerous activities occur outside of 200-mile jurisdictional boundaries, including scientific research, fishing, and international navigation (both commercial and military). Whether the Authority would play a role in any of the above activities, or have any jurisdiction whatsoever over activities taking place in the water column beyond national jurisdiction, has been a matter of concern. It is not clear whether Ambassador Pardo, in his speech of November 1967 before the First Committee of the United Nations General Assembly, advocated the establishment of an international organization to act as a trustee over simply the Area, or more comprehensively, the Area and its superjacent waters. Later, in the working paper he prepared for the government of Malta in 1971, he called for a new comprehensive legal order for the deep oceans encompassing not only the seabed but the water column as well.[18] Entitled a "Draft Ocean Space Treaty," the document claimed that ocean space is a single and distinct ecological system in which there are crucial links among all ocean space activities. As a consequence, international management of all ocean space beyond national jurisdiction, not just the seabed, would be required to adequately maintain the system.

Such ideas were anathema to the major maritime powers who viewed freedom of navigation as a nonnegotiable item during UNCLOS proceedings. Without freedom of navigation throughout the vast reaches of the oceans, the ability of the superpowers to maintain hegemonic domination over global affairs would be at an end. During the pre-1970 Seabed Committee debate, many delegates from developing nations expressed their support for a comprehensive ocean regime.[19] Even during the Caracas UNCLOS session there were calls for international control over ocean space, e.g., the statement of the Mexican delegate: "The new regime should not be limited to exploration and exploitation of the seabed; ocean space must be considered as a whole, and the future authority should administer all resources in the seabed as well as the water above."[20] Sufficient political will to advance its evolution, however, was obviously lacking, as the 1970 Declaration of Principles states that although the seabed beyond national jurisdiction is the common heritage of mankind, nothing in the principles shall affect "the legal status of the waters superjacent to the Area or that of the air space above these waters."

Friedheim and Durch contend that since 1970 there has been very little support for international control over activities in the water column superjacent to the Area.[21] Much to the relief of the major maritime powers, negotiation to retain traditional high-seas freedoms beyond national jurisdiction has not been necessary. All the negotiating texts from the SNT through the ICNT have contained an article clearly stating the following:[22]

> Neither the provisions of the Convention nor any rights granted or exercised pursuant thereto shall affect the legal status of the waters superjacent to the Area or that of the airspace above these waters.

The possibility of the international community endowing the Authority with regulatory responsibilities for activities in the water column at some future time, however, remains a serious concern of the maritime powers. One cannot state with assurance that sentiment for greater international control over ocean space beyond national jurisdiction will remain dormant. Quite the opposite could occur, for example, if the vast majority of the international community judged the Authority a success over time.

The Soviet position toward the Authority has, from the beginning, been largely influenced by the Soviet concern over potential encroachment of the Authority upon traditional high-seas freedoms.[23] As a basically self-sufficient nation, in terms of minerals, the Soviets are less concerned than other nations with seabed mining per se. The Soviet proposal at Geneva to split the Area between Authority jurisdiction and that left to states was widely viewed by the Group of 77 as an attempt to ensure that wide expanses of the oceans would not ultimately come under the control, or at least be affected by, the Authority.[24]

U.S. opposition to a strong supranational Authority is undoubtedly influenced by similar concerns. It is difficult to assess what weight to give them in the formation of domestic policy because of the strong mining interests pushing in a similar direction. Military officials are content to maintain a low profile while letting mining interests argue the U.S. position most forcefully.

Whether there will be persistent international pressure throughout this century for "creeping vertical jurisdiction" over ocean space beyond national jurisdiction is a matter for speculation. We have already witnessed since World War II a strong and consistent enclosure movement extending national jurisdiction into far offshore areas hitherto free from national regulation. Whether a seabed Authority will at some time in the future demonstrate a similar willingness to extend its jurisdiction (upward not horizontally) must remain a latent U.S. concern.

Conclusion

The chances of producing at UNCLOS basic conditions for mineral exploration and exploitation that are acceptable to existing American-based mining organizations must be viewed as problematic. The task of carefully formulating mining specifics would be both difficult and time-consuming even if common values prevailed. Without agreement on fundamental premises, it becomes nearly impossible. Clearly, a mutually satisfactory interdependent relationship between the Authority and mining groups must be forged. Mining groups need the Authority to gain international legitimacy for their endeavor. The Authority, on the other hand needs mining groups to provide the expertise and technology necessary to carry out its mandate effectively.

Notes

1. Jack N. Barkenbus and Dennis Pirages, "The Future of Mineral Interdependence," Paper presented at the American Political Science Association Meeting, San Francisco, Calif., Sept. 1975.
2. Working with the Organization of Petroleum Exporting Countries (OPEC) has been profitable indeed for the major oil companies.
3. Elisabeth Mann Borgese, "Shaping the Law of the Sea," *World Issues,* 2, 4 (Oct./Nov. 1977), p. 17.
4. Roy Skwang Lee, "Machinery for Seabed Mining: Some Issues Before the Conference," in F. Christy, Jr., *et al.,* eds. *Law of the Sea: Caracas and Beyond* (Cambridge, Mass.: Ballinger, 1975), p. 122.
5. The document is titled "Proposals Regarding Conditions of Exploration and Exploitation" and is a comparison of the conditions stated in the following major proposals at the Caracas session: The United States (A/Conf. 62/c.1/L.6), the Group of 77 (A/Conf. 62/c.1/L.7), the European 8 powers, (A/Conf. 62/c.1/L.8), Japan (A/Conf. 62/c.1/L.9). The document was originally issued August 23, 1974, as CP/Working Paper No. 2, and reissued March 18, 1975, at the Geneva session.
6. "Opening Statement by C. W. Pinto," CP/Cab. 11 (c.1), March 19, 1975.
7. Testimony of John Norton Moore, in *Status Report on Law of the Sea Conference, Part 3,* Subcommittee on Minerals, Materials and Fuels, 94th Congress, 1st Session, June 4, 1975, p. 1214.
8. A/Conf. 62/c.1/L.8.
9. R. L. Friedheim and W. J. Durch, "The International Seabed Resources Agency Negotiations and the New International Economic Order," Paper delivered at the American Political Science Association, Chicago, Ill., Sept. 2–5, 1976, p. 10.
10. SNT, Art. 30(2b).
11. Speech by Secretary of State Henry A. Kissinger, "The Law of the Sea: A Test of International Cooperation," New York, April 8, 1976.
12. SNT, Art. 10(3c).
13. For a detailed examination of the controversy surrounding scientific research in the oceans, see Warren Wooster, ed., *Freedom of Oceanic Research* (New York: Crane and Russack, 1973).
14. RSNT, Art. 10(1).
15. RSNT, Art. 10(3c).
16. Kissinger's 1976 address has already been referred to in note 11. In 1975 he addressed the American Bar Association on "International Law, World Order and Human Progress," Montreal, August 11, 1975.
17. *Status Report on Law of the Sea Conference, Part 5,* June 8, 1976, p. 1623.
18. A/AC. 138/53.
19. A/8021/Annex 3.

20. U.N. press release, SEA/72, July 17, 1974.
21. Friedheim and Durch, "The International Seabed Resources Agency Negotiations and the New International Economic Order," p. 17.
22. SNT, Art. 15; RSNT, Art. 15; ICNT, Art. 135.
23. Mark W. Janis and Donald C. F. Daniel, *The USSR: Ocean Use and Ocean Law,* Occassional Paper # 21, Law of the Sea Institute, May 1974, p. 16.
24. Edward Miles, "An interpretation of the Geneva Proceedings–Part I," *Ocean Development and International Law* (Spring, 1976), p. 32.

PART 3

Contrasting Visions of the Future

Chapter 9

U.S. Negotiating Strategy

Up to this point the specific areas of conflict between the North and South over deep seabed mining have been discussed, e.g., price and production controls, an acceptable system for mining, establishment of the Authority's decision-making organs. Very little has been said, however, regarding the "hidden agenda" of this dispute and more specifically why intransigence has characterized the many years of UNCLOS negotiations. To be sure, access to manganese nodules, in itself, is deemed an essential goal of industrial nations; yet there is much more at stake than simply minerals. The intransigence of the South, on its face, is even more difficult to explain since so few developing nations have significant substantive interests in deep seabed minerals. Clearly, one must go beyond the issue at hand to understand the very real dynamics of the dispute.

We have seen that the issue of deep seabed mining has been with us for over a decade. Only since the 1975 Geneva UNCLOS session, however, has it been singled out as the major obstacle to concluding a comprehensive treaty on the law of the sea. Given the multitude of difficult, sensitive, and significant issues under UNCLOS consideration, it is surprising that the deep seabed mining dispute would be labeled the prime impediment to a "package" accord. One would think that issues of such high economic and strategic salience as territorial limits, passage on international straits, or fishing disputes would prove the most difficult ones to negotiate. The purpose of this section will be to explore and explain why negotiations over the formation of a new international organization to regulate deep seabed resources have progressed so unsatisfactorily.

To understand why this issue remains so difficult to resolve one must recognize its enormous symbolic significance. There is scarcely an issue before the international community today that so clearly and cleanly splits along North–South lines. Although the developing nations support OPEC ideologically, their animosity at having to pay full OPEC prices is well known. One need not go outside the UNCLOS forum, however, to observe divisions within the developing world; witness the dispute between land-locked and coastal states over the extension of national boundaries into ocean space. On the deep seabed issue however, there are no strongly entrenched countervailing interests among the developing world. There is no significant split between producers and consumers that would inhibit them from uniting and pressing for larger third-world goals.

To the North, and particularly the United States, the manganese nodule issue as a symbol, far transcends that of resources. Growing resource dependency, the changing and diminishing role of MNCs, and concern over the future role of the United States in international organizations all come together on this issue and produce a policy response that at first appears incongruous with the specific issue under consideration. For policy analysts it is tempting to attribute the hard line of the United States to the influence of the mineral industry, and one can make a case for this point of view. Officials from mining entities have been prominent among the large advisory body established by the U.S. UNCLOS delegation. One sign of the mineral industry's strength in policy formation is the fact that for a number of years the chief U.S. negotiator in Committee I was an Interior Department lawyer and not a State Department representative. When this responsibility was turned over to the State Department in 1977, the former chief negotiator joined a law firm that subsequently represented Kennecott Copper.[1] A very strong case can be made, therefore, that the U.S. position in Committee I has been the result of industry interests and lobbying.

Students of public policy who perceive policy formation as a power struggle among bureaucratic units can also make a strong case. One might expect an Interior Department versus State Department contest for dominance, and this dimension is evident. Perhaps the most significant development in bureaucratic politics, however, was the entrance of the Treasury Department into policy formation on the law of the sea during 1973. This department weighed in heavily against any U.S. policy that would, in effect, exchange resources and economic efficiency for political gains. The Treasury Department insisted that a strong regulatory body, such as an international Authority, was not only unnecessary for deep seabed mining but harmful to the long-range interests of the American consumer. The entrance of the Treasury Department into the formation of Committee I policy marked a new stage in U.S. deliberations.

Both the group interests and bureaucratic approaches to U.S. policy-making are revealing; nonetheless, they do not capture the totality of the issue, or more important, what the author would term its essence.

No other issue at UNCLOS strikes so close to the larger struggle over global resources between North and South than this one; consequently, what is in dispute is not only deep seabed mining, but more importantly, alternative visions of a future economic and political international order. It is because national decision-makers cling so tenaciously to their preferred (and conflicting) visions of the future that accommodation has been so difficult to achieve. The preferred U.S. vision of the future, not surprisingly, is one that retains U.S., or at least Western, hegemony over global economic transactions. This desire for hegemony, often cloaked in more technical terms like "economic efficiency" and "free-market assurances," has led directly to U.S. insistence on a strong MNC role in deep seabed mining. The developing nations, on the other hand, prefer the demise of U.S. and Western economic hegemony and the deliberate structuring of a more symmetrical relationship among nations. A necessary component of such a future would be the subordination of MNCs to the dictates of national or international control.

One might conclude that seabed mining is, in itself, a rather innocuous and even insignificant arena in which to carry on global battles. Nevertheless, the fact is that there is considerable fear or hope, depending upon the viewpoint, over the precedent that would be set by the establishment of the Authority. It is generally recognized in this country and in others that a new seabed Authority could well become a harbinger of a new era. The potential "domino effect," if you will, of such an Authority upon the structure or restructure of other international organizations could present problems for U.S. policy-makers in other forums and strengthen the resolve of the developing nations to forge a new world order. It is not surprising then, when viewed in this light, that such an extended struggle should attend the Authority's formation.

Since the beginning of this decade, international organizations increasingly have been the forums for persistent international power struggles over political and economic control. These forums provide a particularly propitious environment for the developing nations to do battle because they have the advantage of numbers and collectively have the power to challenge the existing order. What is involved in the formation of an international organization today is not simply a response to functional imperatives but a test of bloc strengths. Anyone approaching the question of Authority formation by assessing what sort of organization is *necessary* to do the job (i.e., seabed mining), misses the whole point of the negotiations. The essential purpose has been not to design a smoothly functioning international Authority but to form an Authority indicative of existing and future power alignments. Manganese nodules and the prospective miners have, therefore, become ensnarled in a much larger issue.

The intransigence of the U.S. position at UNCLOS was noted previously. Despite the fears of interested mining groups that U.S. negotiators would "sell out" their interests, no such development has taken place. There are those who claim that the United States has already made major concessions during recent UNCLOS sessions. However, when looking carefully at these so-called concessions, one finds more movement than substance.

The major change in the U.S. position has been its verbal support for Enterprise operations. Henry Kissinger in his April 8, 1976, speech stated, "The United States could accept, as part of an overall settlement, a system in which prime mining sites are reserved for exclusive exploitation by the Enterprise or by the developing countries directly." This offer, however, was predicated upon acceptance of the banking system; in the same speech Kissinger stated, "What the United States *cannot* accept is that the right of access to seabed minerals be given exclusively to an international authority or be so severely restricted as effectively to deny access to the firms of any individual nation, including our own."[2]

The banking system, however, is little more than an ingenious formula to escape international control by splitting the Area. Favorable manganese nodule sites are likely to be so abundant that mining companies can readily afford to simply give away half, and the banking system allows industrial nations to obtain all the seabed sites they could possibly ever use. Whatever "concession" there might be in this system, therefore, is illusory. Although it appears that the banking system is an appropriate "middle ground" between the licensing and the pure Enterprise alternatives, it is hardly that in practice. The Enterprise, under such a system, is likely to be and remain an empty shell, unable to muster the expertise and technology necessary to effectively carry out its assigned tasks. According to the ICNT, the Enterprise would have a governing board elected primarily on the basis of equitable geographical distribution–which would mean that it would consist primarily of developing nations with little or no experience in mining operations. With their own mining sites secured, mining companies are likely to either ignore the Enterprise or bring less than a maximum amount of diligence to a joint venture. Under such circumstances, it is unlikely that the Enterprise would ever be able to mine competitively, and the worthiness of the entire Authority would be called into question. During the 1977 UNCLOS session, the United States reportedly offered substantial funds to launch the Enterprise's first mining venture (contingent upon acceptance of the banking system, of course). When developing nations insisted upon firm assurance of technology transfer to the Enterprise, as seen in the ICNT, the United States balked. In itself, therefore, the banking system simply fails to ensure the satisfactory establishment of a viable international organization.

Another U.S. proposal made during the 1976 UNCLOS session was for a temporary limitation (for a period of fifteen to twenty years) on the production of seabed minerals tied to the projected growth in the world nickel market, which was estimated to be about 6 percent a year. This proposal caused much consternation amoung U.S. mining officials, particularly when the government had, throughout the early 1970s, insisted that the United States would never agree to a treaty that included production controls. On the face of the proposal, therefore, it appeared to be a major concession to the developing nations that were land-based mineral producers. In fact, however, there was little chance that production controls would ever be implemented under such a scheme. A U.N. study indicated that this proposal, if implemented, would have authorized twice as much mining capacity as was being planned for the first decade of operations.[3] The possibility of nodule production reaching the established limits, therefore, would not begin to appear until it was time for such controls to be lifted. U.S. mining company officials obviously were aware of this fact, and their protest was based not so much on the specific proposal but upon the precedent that was set and the fear that it might lead to the imposition of real controls. As reflected in the 1977 ICNT, developing nations went much further in proposing production controls which were renounced by the United States. We can see, therefore, that U.S. officials in 1976 came out in favor of production controls because they realized that under their scheme it was unlikely that they would ever go into effect. Production controls in the strict sense of the words were another matter.

One must, in a certain sense, admire these U.S. schemes, which appear to offer so much while at the same time give so little. The appearance of accommodation has certainly influenced the major U.S. newpapers, whose editorials after each inconclusive UNCLOS session point to G-77 intransigence as the major negative force in the negotiations. Those UNCLOS delegates who understand the subject, however, are not so easily deceived. What we have found is that there has been no significant concession in U.S. policy during the decade of the 1970s (when "concession" is used in the strictest sense of the word—"to yield" or "to concede"). What U.S. negotiators have sought over the years is international sanctioning of their preferred plans—which is a far cry from seeking an international resolution of the issue.

U.S. Strategy

As stated previously, U.S. intransigence over the seabed issue stems from not only its mineral interests per se but also the perceived imperative of retaining hegemony over global economic transactions. U.S. policy-makers are unwilling to make significant concessions to the developing world over this issue; moreover, they have persisted in believing that such concessions are unnecessary. This section discusses five distinct premises that have guided U.S. policy-makers over the years of negotiation and led them to believe that political compromise was neither necessary nor desirable: (1) a sense that coastal developing nations would trade the more "ephemeral" political and ideological goals sought in UNCLOS Committee I for the hard economic rewards that could be gained in Committee II negotiations; (2) a sense that Committee I negotiations were being stalled, not over political disputes, but rather by the substantive economic interests of a few mineral-exporting developing nations; (3) a sense that the manifestation of strong congressional pressure for unilateral mining legislation would lead developing nations to accede to the U.S. position; (4) a sense that if high level government officials could intercede and thereby demonstrate to other nations the seriousness with which the United States approached the negotiations, the deadlock would be broken; (5) a sense that ultimately the G-77 had no choice but to concede since the United States and other Western nations were the sole possessors of the necessary technology.

In short, U.S. policy-makers assumed that the combination of the foregoing factors would lead the G-77 to accept the U.S. position. This assumption was incorrect. Although the failure of this strategy is now evident, it is worth examining in detail the premises upon which it was based.

NATIONAL VERSUS INTERNATIONAL GOALS

In private discussions, the opinion has often been expressed that Committee II and not Committee I is the more important. This belief stems from the predominant focus of coastal nations throughout UNCLOS sessions upon the extension of national boundaries to enclose the vast resources of the ocean. Indeed, if one were to characterize UNCLOS negotiations as a whole, the consensus for nationalization of ocean space out to 200 miles has to be the major development. In general, U.S. policy-makers felt that if exchanges were to be made among the various issues being negotiated, the G-77 would view Committee I issues as more expendable than the economically substantive issues of Committee II. Few believed that the G-77 would risk sacrificing the economic gains made in Committee II for the broad, ideological goals expressed in Committee I. As was shown in Chapter Five, Latin-American devotion to the Enterprise was initially conceived of as a bargaining chip, much as the United States envisioned it. Only later did it come to be a G-77 goal in its own right. The expected trade-offs between Committee I and Committee II never materialized, therefore as U.S. policy-makers had planned.

MINERAL-EXPORTING DEVELOPING NATIONS

The second mistaken premise of U.S. policy was the belief that only a small band of land-based mineral producers among the G-77 were obstructing progress in Committee I. In particular, it was thought that the copper-exporting nations, notably Peru and Chile, were the chief antagonists. Their purpose in stalling negotiations was evident—a fear that imminent seabed mining would seriously erode their position as exporters and result in grave problems of foreign exchange. U.S. officials took the opposition of these countries seriously, for obvious substantive economic interests were at stake. It was also deemed significant that the Peruvian delegate held an important position in the negotiations (as the coordinator for the Committee I contact group of the larger G-77). The Caracas UNCLOS session may have contributed substantially to overestimating the power of land-based mineral producers because the negotiations focused on the economic implications of deep seabed mining for these producers.

As mentioned previously, Secretary of State Kissinger on April 8, 1976, announced that the United States was willing to place temporary limits on mineral production from the seabed. The proposal was eagerly accepted by the G-77 mineral exporters and elaborated upon in the RSNT. Nevertheless, the intended effect of the U.S. compromise was never realized. Although several G-77 mineral exporters were now willing to negotiate a complete treaty, they could not convince their G-77 colleagues to be equally enthusiastic—simply because the United States presented no political concessions for the vast majority of G-77 nations equivalent to the economic ones offered to the mineral producers. The futility of this strategy is apparent from the continuing negotiating deadlock. Clearly, it was simplistic of U.S. policy-makers to believe that a few mineral-exporting nations could dominate the thinking of approximately one hundred other nations. Interestingly, the delegate from Peru had publically stated that G-77 mineral producers had no dominance over the G-77 position.[4] U.S. policy-makers apparently preferred not to believe him.

CONGRESS AND THE USE OF PRESSURE AS A STRATEGY

Although the initiative in ocean policy-making lies with the executive, various congressional committees have consistently made their preferences known. This attitude is proper, for the Senate ultimately must approve a law-of-the-sea treaty, and both branches of Congress will have to pass implementing legislation. The congressional figure most prominent on deep seabed mining issues was before his death in 1978, Senator Lee Metcalf. Beginning in the early 1970s and continuing throughout this decade, Senator Metcalf held hearings on seabed mining bills that, if passed, would have bypassed UNCLOS negotiations and permitted the domestic U.S. mining firms to begin seabed exploitation.

Senator Metcalf, as well as many other observers, felt that by raising the specter of unilateral U.S. licensing, the negotiations at UNCLOS would proceed more expeditiously. In other words, it was claimed that such pressure would act as a spur in bringing other nations around to the U.S. position. The Metcalf legislation, therefore, was viewed as a potentially valuable catalyst in promoting agreement or in encouraging other nations to accept the U.S. position. One observer of this strategy noted, "Negotiators gain certain bargaining advantages internationally from having this 'club in the closet.' Foreign delegations know there is a segment of the U.S. Government that wishes to proceed with unilateral solutions to what it views as pressing problems and that has no faith in the international negotiating process."[5]

The "club" was virtually brought out of the closet during the first week of the initial 1976 New York UNCLOS session. During that period, Senator Metcalf had his bill reported to the Senate Committee on Interior and Insular Affairs. A similar bill sponsored by Congressman John M. Murphy, chairperson of the House Oceanography Subcommittee, was at the same time reported to the House Merchant Marine Fisheries Committee. Congressman Murphy went even further, issuing a press release in which he labeled UNCLOS a "sham" and stated his intention of moving ahead with unilateral mining legislation.

These congressional pressure tactics may have influenced some G-77 delegates, but not enough to produce the desired result. The fact that congressional pressure did not have a favorable influence on UNCLOS delegates in 1976 did not cause the administration to seek a new substantive basis for accommodation, but instead led it to join forces with Congress in subsequent years in seeking U.S. legislation.

These pressure tactics, instead of producing the desired result, have contributed substantially to the polarization of the issue and have worked directly into the hands of those G-77 members most ideologically committed. Again, the mistaken perception by U.S. policy-makers of the G-77 position led them to believe that pressure could be effective.

HIGH-LEVEL INVOLVEMENT

The lack of high-level administration attention to ocean issues is seen as a mistake by several observers, including Senator Claiborne Pell. Senator Pell has long urged the administration to place either the vice-president or the secretary of state in direct charge of UNCLOS negotiations to demonstrate to other nations the seriousness and importance the United States attaches to them. Members of the U.S. negotiating team for many years urged Secretary of State Kissinger to take an active role in the negotiations. However, until the 1976 sessions he resisted.

In his New York speech on April 8, Kissinger assessed in some depth the nature of UNCLOS negotiations and made proposals to move them forward. The speech was a marvelous example of offering carrots with one hand while holding a stick in the other. The carrot, offered to mineral-producing developing nations, was a temporary limit on seabed mineral production tied to the projected growth in the world nickel market. The stick, intended to have an effect upon all nations, was the threat to move ahead unilaterally:

> We strongly prefer an international agreement to provide a stable legal environment before [mining] development begins, one that ensures that all resources are managed for the good of the global community and that all can participate. But if an agreement is not reached this year it will be increasingly difficult to resist pressure to proceed unilaterally.[6]

Kissinger's message to foreign delegates, emphasizing that the United States would not wait much longer for a seabed treaty, was clear even if couched in conciliatory language. The essential purpose of his speech was to make the threat of unilateral action more credible. It was thought that a threat made by the highest U.S. foreign policy official would have a greater impact upon delegates than one delivered by U.S. UNCLOS negotiators. Moreover, Kissinger announced that President Ford had asked him to head the negotiating team during a second 1976 UNCLOS session—which was intended to demonstrate a new seriousness attached to UNCLOS negotiations by the President. Kissinger's entrance into UNCLOS negotiations did not portend any far-reaching shift in U.S. policy, however; rather it signalled a new importance attached to producing *the desired* seabed treaty.

The impact of Kissinger's involvement in UNCLOS negotiations was negligible. The carrots were distributed to too few G-77 nations, and the obvious stick did not frighten the majority of delegates. In fact, Kissinger's participation in the two 1976 New York sessions was extremely limited, amounting to perhaps four full days of negotiating. Because U.S. policy remained basically the same during the sessions, no discernible progress was made in reaching an international consensus.

Ironically, just when Kissinger entered the negotiations to add credibility to the U.S. position, his own credibility was undermined by the presidential elections. Foreign delegates, suspecting that a new administration might adopt a new seabed policy, were unwilling to make compromises with a secretary of state whose term of office was possibly limited.

UNILATERAL MINING

Perhaps the most compelling reason why the United States has yet to come to grips with G-77 political goals is the belief of many congressmen that ultimately the G-77 has no choice but to accept a regime favored by the United States and other industrial nations. This belief appears to be justified, since the industrial nations have the technology and expertise to conduct actual mining, and the G-77 could presumably find themselves locked out of the mining enterprise altogether. Senator Hollings of South Carolina articulated the sentiments of several congressional colleagues when he stated, "Since the U.S. possesses the technology to develop the resources of the deep ocean, it is in a strong bargaining position. It should not give it up easily."[7]

U.S. negotiators at UNCLOS have been inhibited from striking the necessary bargain with G-77 nations in large part by fear of the congressional furor that might result. It is well to keep in mind that the Senate must ratify all treaties by a *two-thirds* majority. Because of the highly salient symbolic nature of the issue, it is not difficult to imagine how a large number of senators might respond to exhortations decrying a U.S. negotiating "sell-out" on this issue. The administration has had to face this serious constraint throughout every UNCLOS session.

It is a widely held misconception on Capitol Hill that a unilateral U.S. sanction of mining would occasion few, if any, political costs. What is not properly appreciated is the degree to which the substantial U.S. ocean activities are dependent upon a stable international climate in general, and a uniform sea law in particular, and the range of actions G-77 nations could conceivably take in retaliation.

Because nations are so polarized over this issue, international reaction to unilateral mining is likely to be intensely negative and vociferous. At the very minimum, there would be a legal test of U.S. authority to unilaterally mine what the international community (including the United States), in a 1970 U.N. resolution, had declared to be the common heritage of mankind. While the case was being brought before the International Court of Justice or debated within the United Nations, it is likely that developing nations would establish an Authority, which they would claim, represents the only legitimate body to conduct exploitation of the deep seabed.

All these reactions can be ignored by the United States. What cannot be ignored, however, are the potential acts of terrorism which might be aimed directly at the mining vessels themselves or at downstream processing facilities. Security at sea would require that military protection accompany each mining ship as well as those vessels transporting manganese nodules to shore for processing.

Retaliation for unilateral mining, however, need not be directed solely at the mining enterprises themselves. It is very possible that unilateral mining would lead to the collapse of UNCLOS negotiations—and consequently all the compromises already reached on vital issues, such as territorial seas and resource zones, commercial and military passage through straits, and marine pollution, would be placed in jeopardy. U.S. negotiators have expressed general satisfaction over the course of these negotiations. It would definitely not be in the national interest, therefore, to proceed with mining if it would prevent important negotiated issues from becoming treaty laws.

Unilateral mining need not occasion solely a marine-related response. There is a North–South confrontation on a growing web of issues, of which sea law is a small strand, and a tug on this particular strand could effect others. Despite the South's obvious economic and military deficiencies vis-à-vis the North, it does have potential leverage.

Conclusion

The thesis of this chapter has been that the failure of U.S. policy-makers to seriously address the political goals of the G-77 has substantially contributed to the deadlock in current UNCLOS negotiations. Clearly no amount of negotiating pressure will force G-77 nations to adopt the vision of a seabed regime preferred by the United States. Equally as clear is the fact that if the United States seriously desires an international solution to the seabed dispute, it will have to make political as well as economic compromises. Admittedly, it is much more difficult for U.S. policy-makers to make political rather than economic concessions without jeopardizing basic national interests. If the negotiations were a question of creating formulas to divide economic resources, consensus would have been obtained long ago; but when polarization occurs, political control is less divisible.

Notes

1. Deborah Shapley, "Ocean Mining: Former Negotiator Now Lobbies for Kennecott," *Science,* 196, 4293 (May 27, 1977), 964–65.
2. In *Status Report on Law of the Sea,* Part 5, June 8, 1976, pp. 1940–47.
3. *Economic Implications of Sea-Bed Mining in the International Area: Report of the Secretary General* (A/Conf. 62/37), United Nations, Feb. 18, 1975, p. 14.
4. Statement by Alvaro de Soto, in F. Christy, Jr., *et al.,* eds., *Law of the Sea: Caracas and Beyond* (Cambridge, Mass.: Ballinger, 1975), p. 157.
5. Kenneth Kolb, "Congress and the Ocean Policy Process" *Ocean Development and International Law,* 3, 3 (1976), 277.
6. *Status Report on Law of the Sea,* p. 1946.
7. Statement of Senator Ernest F. Hollings, in *Perspectives on Ocean Policy,* NSF, Ocean Policy Project (Baltimore: Johns Hopkins University Press, 1974), p. 373.

Chapter 10

The New International Economic Order

Major structural shocks or changes in the nature of the international economic system are certain to be reflected in or affect sea law negotiations. It is the author's belief that events occurring outside of the Seabed Committee and UNCLOS were the determining factors in the increasing polarization among nations within Committee I. More specifically, it was the petroleum embargo of late 1973 (and subsequent petroleum price raises instigated by OPEC), and the call for a New International Economic Order (NIEO) that led to Committee I polarization.

This chapter is devoted to an analysis of how events associated with the larger international economic system have affected the South's unified position at UNCLOS. It is claimed that the lack of substantive interest in manganese nodules by developing nations has made their ideological unity possible, and that without good-faith proposals for compromise by the North this unity will be difficult to crack.

Seabed Committee

From the beginning of Seabed Committee deliberations in the late 1960s, major differences between North and South nations over what constitutes the proper ocean regime have been evident. In a paper prepared in 1969, Friedheim[1] noted the differing phrases often used in the speeches of North and South delegates to the Seabed Committee (see Table 15).

As was pointed out earlier, few developing nations were optimistic that their preferred visions of an appropriate seabed regime would be realized.[2] Industrial nations had dominated rule-making during the preceding 1958 and 1960 conferences, and despite their greater numbers in the late 1960s and early 1970s there appeared to be little reason to believe that developing nations would be any more successful than in previous years. Few observers in the early 1970s predicted that deep seabed mining would be the stumbling block in UNCLOS negotiations.

TABLE 15. Key Phrases in the Seabed Debate

SOUTH/LESS DEVELOPED COUNTRIES' PREFERENCES	NORTH/DEVELOPED COUNTRIES' PREFERENCES
Protect the rights of coastal states	Protect freedom of the high seas
Protect the economies of developing states	Protect fishing rights
Prevent exploitation by technologically advanced states	Protect freedom of scientific research
Prevent colonialism, imperialism	Protect the rights of all
Close gap between developed and developing states	Protect the access of all
Strengthen ocean capabilities of developing states	Protect maritime interests
Taking into account special needs of developing states	Take into account international law
Rights to sovereignty or exploitation not implied by scientific research	Take into account U.N. Charter
	Take into account existing treaties

SOURCE: Robert L. Friedheim, "The Marine Commission's Deep Seabed Proposals—A Political Analysis," in *The Law of the Sea: National Policy Recommendations,* 4th Annual Law of the Sea Institute Conference, June 23–26, 1969, p. 91.

However, at this time, a significant perceptual shift occurred among developing nations that was to have a significant impact upon negotiations. When the boundaries separating national from international waters were in question, much of the interest of developing nations in the seabed was economic in nature. There was considerable hope that significant revenues could be raised through mining the seabed and distributed among the developing nations. The insistence of the coastal nations, however, upon extended 200-mile EEZs eliminated this prospect. Rather than dropping interest in the deep seabed, however, developing nations shifted to a concern over the role they would play in the actual exploitation of the Area itself. To recall Principle I of the Declaration of Principles, it is explicitly stated that *both* the Area and its resources are the common heritage of mankind. Developing nations came to insist that only through their participation in the management of the Area could resource exploitation, regardless of its magnitude and economic rewards, be regulated on behalf of the international community.

This shift in focus or perception was a significant development in the evolution of negotiations, but it alone was not a sufficient cause for deadlock at UNCLOS. Louis Sohn of Harvard, who has been actively involved in UNCLOS from early Seabed Committee deliberations, has stated that North and South were slowly moving toward accommodation during the early 1970s until late 1973 and 1974.[3] It is worth exploring what produced the deadlock.

Committee I deliberations were, I submit, affected crucially be events that occurred outside the confines of UNCLOS during the years 1973 to 1975. First, the Arab oil embargo of 1973–74 and the subsequent OPEC price hike changed the fundamental context within which seabed mining had been viewed up to that time. Before 1974, the U.S. government was generally supportive of domestic miners who wished to embark on this new commercial endeavor, as such mining would add another source of minerals. The context shifted after 1974, however, as the commercial aspects of the enterprise began to be downplayed and their strategic importance "discovered."

Although the mineral industry before 1974 was certainly aware of the changing climate for mineral exploitation globally and the growing precariousness of their investments, this fact did not register publicly until the events of 1973–74. The OPEC price hike and Arab embargo forcefully demonstrated to the industrial states the vulnerability that accompanies resource dependency. Up until this time, the dangers involved in importing vital industrial minerals were generally overlooked, or at least viewed as potential costs that did not outweigh the benefits. In the wake of OPEC action, manganese nodule mining became inextricably linked with fears over excessive dependence (as described in Chapter Four). The transformation, in the minds of many U.S. policy-makers, of nodules from commercial to strategic entities has brought a heightened dedication to conclude a treaty resting squarely within the U.S. mineral interests.

If the petroleum embargo and OPEC price hike had an important influence upon the U.S. resolve within Committee I, it had an even more powerful effect upon the perceptions and interests of developing nations. The important lesson gained by the South was that developing nations could indeed have a powerful impact upon the international economy if sufficient unity and political will were brought to bear. This was a striking revelation to nations whose economies have historically been tied to actions and events initiated by the major colonial or developed powers.[4] Developing nations perceived for the very first time that economic power throughout the world was indeed shifting and that they now had the opportunity to become more than simply passive recipients of economic policy. In terms of the seabed, it meant that developing nations no longer felt they need defer to the wishes of industrial nations—as they had expected would be the case at the beginning of the decade.

The events of 1973–74, therefore, produced responses at UNCLOS that went in diametrically opposite directions. Were these events perceived of as transitory phenomena, the lines would not have been drawn so sharply and the polarization would have evaporated over time. However, the oil-related events were viewed as a harbinger of times to come. Barraclough has stated, "In the wider perspective of history, it may well turn out that the long-term significance of the 'oil crisis' is the way it has served as a catalyst for the wider and more fundamental confrontation between the poor nations and the rich, which threatens to engulf the world."[5] Polarization at UNCLOS, therefore, must be viewed in terms of what OPEC has wrought.

A New International Economic Order

The period from 1974 through 1975 was an active one for developing nations in the formulation of a preferred, collective vision of the future. This effort took place across the globe, centered at the United Nations but also in meetings at Dakar, Karachi, and Cocoyoc. There was general consensus among developing nations at these meetings that the existing international economic system had failed to adequately address global economic needs and that this failure was the direct responsibility of the industrial nations. There was also consensus that a new international economic order should take the place of the old as soon as possible. One national representative stated, "Our real choice is between our present system which is largely guided and manipulated for the benefit of the rich countries, and a system directed towards solving the problems of division of income and property, of scarcity of natural resources, and of the despoiled environment."[6]

It should be recognized that many of the elements or proposals outlined in the texts emanating from the 1974–75 meetings, and which constitute the blueprint for a new international order, are not generic to these meetings but have been put forward in previous U.N. forums—most notably during the United Nations Conference on Trade and Development (UNCTAD). These proposals include commodity agreements and price stabilization plans, arrangements allowing for greater market access of third-world products, finance and monetary reform with greater distribution of special drawing rights (SDRs) to developing nations, a heightened dedication toward the successful transfer of technology among nations, and debt relief.

Throughout the calls for a new international order there is a unifying theme, i.e., the need for a consciously planned and managed world economy which would directly help the economies of developing nations. Inherent in this plan is a rejection of the traditional market system as a means of ultimately reducing the income gap between nations. Besides the measures previously listed, new emphasis on the relationships among states and the role of multinational or transnational organizations in the world economy can be found in the Declaration on the Establishment of a New International Economic Order (adopted at the Sixth Special Session of the General Assembly, September 1974).[7] This emphasis was demonstrated by the following principles:

1. the sovereign equality of states
2. full and effective participation of all states in solving global economic problems
3. full and permanent sovereignty of every state over its natural resources and all economic activities
4. regulation and supervision of the activities of transnational corporations
5. preferential and nonreciprocal treatment for developing countries in all possible areas of the international economy

It has been pointed out that what is essentially new about the new international order are not the proposed measures themselves but the tenacity and unity with which they are now being presented.[8] Where once these proposals were offered tentatively as requests, the new militancy of the developing nations now frames them as demands. One representative has stated, "There is an urgent need for the developing countries to change their traditional approach to negotiations with the developed countries hitherto consisting of the presentation of a list of requests to the developed countries and an appeal to their political goodwill which in reality was seldom forthcoming."[9]

The early seabed proposals of developing nations in the late 1960s and early 1970s, therefore, can best be viewed in terms of "requests," which few delegates, even those making them, ever felt would be incorporated into a treaty. The new militancy of the post-1973 years, however, based on a sense of shifting economic power, had its impact on seabed negotiations as the requests of earlier years became firm and unyielding demands.

At this point, the question of why the seabed should become a prime battle ground for achieving a new international order may be asked. Surely developing nations could pick an issue with more substance at stake than manganese nodules; regardless of the outcome, it will hardly result in a significant shift of economic power. However, the answer to this lies not only in understanding the issue's enormous symbolic significance, but also in understanding its nature in the context of UNCLOS proceedings. The third world can adopt a unified position on deep seabed mining unlike other issues at UNCLOS, without injuring the interests of any single developing nation. In contrast, the issue of national boundaries cuts across developing nation interests; coastal nations have consistently favored extended national jurisdictions and land-locked nations have not. Given this split, there is no way that developing nations can, as a group demonstrate the unity on this issue that is sometimes characteristic of others. There are no countervailing interests or pressures within the developing world, however, when it comes to the deep seabed question. Since deep seabed mining has been expected to proceed beyond the limits of national jurisdiction, no national boundaries nor resources therein are immediately at stake.

Were the seabed resources of considerable and immediate significance to numerous nations, the costs of intransigence would be high and compromise would be more likely. With the costs of delay low, however, nations can indulge in ideological dispute—not because so much of economic substance is at stake but because so little is. As one might expect, not all developing nations are equally committed to linking manganese nodules to the new international order. There are numerous, what might be termed "moderate," developing nations who would be willing to come to terms with industrial nations on the seabed issue in order to achieve a comprehensive UNCLOS treaty. Several coastal nations are anxious to ensure the large extension of national jurisdiction that would be accorded to them unambiguously. More "radical" or ideologically committed developing nations at UNCLOS, however,—e.g., Algeria, Tanzania—have been enormously successful within the G-77 negotiating group and have thus far prevailed.

It is the author's belief that were the U.S. negotiating position and stance more flexible, and were the implicit and explicit threats to ignore developing nations eliminated, moderates within the G-77 would gain the upper hand. The heavy-handed U.S. position, however, has played into the hands of the radicals who wish to portray the issue as a stark, old versus new order conflict. U.S. policy-makers, themselves, both in the administration and Congress, are giving the radicals all the ammunition they need to build cohesion within the G-77.

NIEO and the Link with Committee I

Perhaps the crucial text outlining the basis of a new international economic order is the Charter of Economic Rights and Duties of States adopted by the U.N. General Assembly on December 12, 1974.[10] The charter is made up essentially of generalities, as one might expect, but Article 29 explicitly deals with the deep seabed:

> The seabed and ocean floor and the subsoil thereof, beyond the limits of national jurisdiction, as well as the resources of the area, are the common heritage of mankind. On the basis of the principles adopted by the General Assembly in resolution 2749(XXV) of 17 December 1970, all States shall ensure that the exploration of the area and exploitation of its resources are carried out exclusively for peaceful purposes and that the benefits derived therefrom are shared equitably by all States, taking into account the particular interests and needs of developing countries; an international regime applying to the area and its resources and including appropriate international machinery to give effect to its provisions shall be established by an international treaty of a universal character, generally agreed upon.

Thus deep seabed mining is directly linked to the new international economic order through the charter, and at the 1975 Genevea UNCLOS session it became forged in practice. Through the urgings of representatives from such nations as Algeria, Mauritania, and Tanzania,[11] other developing nations began to place the seabed debate in the NIEO context. Committee I Chairman Paul Engo openly acknowledged his debt to the "Declaration on the Establishment of a New International Economic Order" in his formulation of the SNT.[12] It is now worth examining how, specifically, NIEO principles are reflected in the SNT and other UNCLOS negotiating texts.

The SNT proposal for a strong policy-making assembly, where each nation has a single vote, perhaps best illustrates allegiance to the principle of the sovereign equality of states and to an effort to involve *all* states in significant policy participation. The one-nation, one-vote council is another attempt to treat all states equally regardless of their differing technological capabilities and positions of power. Practical implementation of this principle is one of the most difficult elements in the seabed negotiating texts for the industrial nations, and particularly the United States, to accept. U.S. calls for a position of strength on the council, commensurate with its producing and consuming interests in manganese nodules, are ignored in the texts. The fact that industrial nations with technical sophistication have, in the past, been accorded preponderent voting weight on other technical international organizations is also not reflected in the texts emanating from UNCLOS. Hence, should the decision-making organs of the Authority be structured according to the ICNT formula, it would be a significant discontinuity from the way functional international organizations have been structured in the past.

Provisions for the application of price and/or production controls in all the negotiating texts also reflect NIEO emphasis on international control over commodity trade. Whereas the RSNT called for production controls over only an interim period, the ICNT wanted eventually a commodity pact between seabed and land-based producers and consumers that would stabilize both price and production. The pace of seabed mining, therefore, would not be determined by mining enterprises themselves but by the appropriate Authority organ in association with land-based mining interests. Regardless of the system of exploitation ultimately set forth, this plan would allow for considerable international control over the operation of transnational organizations.

Finally, there are many references throughout the negotiating texts to how special consideration in the activities of the Authority should be devoted to the needs and interests of developing countries. In other words, these texts do not assume that third-world needs and interests are furthered simply as a function of mining per se.

Conclusion

A Mexican representative familiar with sea law proceedings stated in 1973, "The principle of solidarity is going to constitute, I believe, one of the clearest characteristics of the international behavior of developing countries in the forthcoming Conference on the Law of the Sea."[13] This was not to be the case uniformally throughout UNCLOS and on all issues. Given the widely heterogeneous makeup of G-77 and each nation's unique geographical setting in relationship to the sea, it is not surprising that interests should arise on other than a North-South basis. For the reasons given earlier in this chapter, unity within the G-77 on the deep seabed issue has not required the sacrifice of national interests, and thus attempts by the United States to divide and conquer have proved futile.

Just as the primary motivating force for North-South polarization has come from outside the UNCLOS forum, so too could a movement for accommodation spring from actions taken elsewhere. For example, were commodity producers and consumers to agree to a new integrated program for commodities under UNCTAD auspices and the establishment of a large common fund to finance buffer stocks, there might be less ideological posturing at UNCLOS and a greater resolve to settle the issue. Unfortunately, such developments do not seem imminent, as witness the rather disappointing conclusion of the Conference on International Economic Cooperation (CIEC) in 1977.[14] The United States and other industrial nations now appear willing in broad North-South economic dialogues to grant the South a better deal in economic transactions, i.e., greater monetary concessions in a number of areas. What the South is seeking on a global scale and at UNCLOS, however, is a new deal, not simply a better one; and there is no sign of accommodation there.

Notes

1. Robert L. Friedheim, "The Marine Commission's Deep Seabed Proposals—A Political Analysis," *The Law of the Sea: National Policy Recommendations,* 4th Annual Law of the Sea Institute Conference, June 23–26, 1969, p. 91.
2. John Ludvik Løvald, "In Search of an Ocean Regime: The Negotiations in the General Assembly's Seabed Committee," *International Organization* (Summer 1975), p. 690.
3. Personal communications.
4. Many third-world scholars have posited that the *dependence* of the third world on the major economic powers is in the long term destructive. Teotonis dos Santos has stated, "Dependence is a situation in which a certain group of countries have their economy conditioned by the development and expansion of another economy, to which the former is subject. In all cases, the basic situation of dependence leads to a global situation in dependent countries that situates them in backwardness and under the exploitation of the dominant countries." Quoted in Dale L. Johnson, "Dependence and the International System," in James D. Cockcroft *et al.*, eds., *Dependence and Underdevelopment* (New York: Anchor Books, 1972), pp. 71–72.
5. Geoffrey Barraclough, "Wealth and Power: The Politics of Food and Oil," *The New York Review of Books* (August 7, 1975), p. 29.
6. "The Seventh Special Session of the General Assembly: Issues and Background," United Nations, Sept. 1–12, 1975, p. 17.
7. U.N. General Assembly Resolution 3201(S-VI).
8. Charles Pearson, "New International Economic Order: Fact or Fantasy," Paper prepared for the Justice in the World Program, Lima, Peru, July 5, 1975, p. 4.
9. "Seventh Special Session," p. 9.
10. General Assembly Resolution 3281, adopted by a vote of 120 to 6 to 10.
11. Edward Miles, "An Interpretation of the Geneva Proceedings—Part I," *Ocean Development and International Law* (Spring 1976), p. 11.
12. Paul B. Engo, "Introduction to the Single Text, Committee I," in *Status Report on Law of the Sea, Part 3,* June 1975, p. 1271.

13. Jorge A. Vargas, "The General Needs and Interests of Developing States," *The Law of the Sea: Needs and Interests of Developing Nations,* 7th Annual Conference of the Law of the Sea Institute, June 26–29, 1972, p. 17.
14. Jahangir Amuzegar, "A Requiem for the North–South Conference," *Foreign Affairs,* 56, 1 (Oct. 1977), pp. 136–59.

Chapter 11

Conclusions

The first ten chapters have essentially been a description of the dimensions of the deep seabed mining issue: its technical, economic, and legal aspects; its implications for U.S. national interests; its context within the UNCLOS forum; and its symbolic and political value. This concluding chapter is meant to be prescriptive rather than descriptive. One should approach this task cautiously, as negotiations have been proceeding now for over a full decade and one would intuitively think that virtually every avenue for accommodation has been explored. However, intransigence on all sides has meant in practice that negotiations have been proceeding on different tracks. Nations unfortunately have been talking *at* one another in the UNCLOS forum rather than *with* one another.

Haas and Ruggie have noted that the international community still approaches the formation of functional international organizations in terms of a zero-sum game,[1] and negotiations over the Authority have provided a vivid example of this fact. The perception of winners and losers has been ever present, implicitly if not explicitly. As a result, Committee I sessions have constituted more of a surrogate for ocean conflict than a forum for the development of a rational and equitable mining system. Few would deny that acrimonious debate is preferable to open conflict on the high seas; yet for those observers who perceived and continue to perceive UNCLOS sessions as an unprecedented opportunity to bring a fresh, united, and long-term perspective to the ocean frontier, there has been bitter disappointment.[2] Instead of seeking means of managing the ocean commons, nations have preferred to find ways of dividing it. Instead of assessing how management and exploitation of ocean resources can fulfill global needs, the focus of UNCLOS negotiations has been on international accommodation to national needs. The unwillingness of nations to start with a global perspective and then to see how immediate national interests can be accommodated (rather than vice versa) has been central to the deadlock at UNCLOS and must somehow be overcome if an international solution is to be found.

U.S. Fears

In a very real sense, the manganese nodule issue is the economic equivalent in American politics of the Panama Canal. Both issues symbolize the changing nature of international relations and the new role of the United States in global politics. U.S. citizens find it difficult to support the disengagement of our military presence along the Panama Canal, not because the canal continues to be of prime strategic importance, but because this disengagement represents a diminution of the American presence abroad. U.S. citizens, accustomed since World War II to seeing their government play the preeminent role in global politics, are not prepared for a new era in which the United States is placed within, rather than atop, the community of nations.

Just as does the Panama Canal, the deep seabed, in its essence, symbolizes the transformation from one era to another. Post-World War II U.S. economic hegemony is now being seriously challenged on numerous fronts: Talk of imposing trade barriers is rife as imports capture a larger share of the U.S. market; competition in foreign markets is fierce, and U.S. companies have gotten a smaller percentage than in years past; and as we have seen earlier, MNCs are now having to change their *modus operandi* in the face of full-scale government intervention globally. Manganese nodule mining is viewed as a way to escape from these winds of change. Even more significant, the basis for the control of functional international organizations, namely, technical expertise and investment, is now being challenged. As we move toward a more interdependent world, which will require the formation of many international bodies, the United States and other Western nations can no longer expect global accession to their playing the predominate role.

Unlike the Panama Canal issue, the deep seabed mining controversy has yet to "go public." The man on the street is unlikely to know that sea law negotiations are proceeding, let alone grasp all the technical complexities and implications that are involved. Since negotiations have been very low key and high-level national officials have been absent from the proceedings, the media have not yet made it a public issue. An exception has been the occasional writings of *New York Time's* columnist William Safire, who has attempted to stir up public reaction against the international forum. In a recent column he characterized Committee I proceedings as "history's greatest attempted ripoff," and he described the ICNT as "a proposal to seize the treasures of the ocean bottom—which belong to 'all mankind' and not to any governmental authority—and to place the ownership largely in a new bureaucracy, bigger than the U.N., dominated by the dictators of the undeveloped nations."[3]

If and when the controversy surfaces as a prominent political issue, we can expect more comment in the Safire vein, and the parallel with the Panama Canal will become more obvious. In both cases, the Senate will have to ratify treaties by a two-thirds majority, and public outcry against a U.S. retreat from hegemony will make ratification very difficult to achieve.

Dilemma of Developing Nations

Despite brave talk of embarking upon a new and structured international economic order, what order in fact will germinate from the disorder of today is not apparent. It is clear that developing nations can create disorder throughout the globe, but it is also clear that they cannot bring a new order to the international scene by themselves. Concerning the oceans, Pinto of Sri Lanka once stated that developing nations are "heirs now to a fortune that they lack the means to claim."[4] Their lack of collective technical expertise in exploitation will require accommodation with those organizations and nations that do possess the requisite technology. Political will cannot produce technological know-how magically.

There is evidence to suggest that most developing nations recognize their dependence on established organizations and the requirements so imposed. For example, they could have by now marshalled a sufficient number of votes within UNCLOS to establish an Authority in their preferred image. They have not done so because they realize that such an organization would be essentially an empty shell.

It may be useful to speculate briefly about how effective the Authority would be if it were to be structured as outlined in the ICNT. Skolnikoff in his book, *The International Imperatives of Technology,* identifies seven conditions or characteristics that make for an effective international organization:[5] (1) specialized, especially technical, subject matter; (2) a clear, justified, and agreed-upon mission; (3) membership restricted in number and on the basis of interest in the subject; (4) organizational structures that reflect interest, power, and knowledge of member governments; (5) a small secretariat; (6) little public interest; and (7) subject matter of moderate political or economic interest.

The Authority described in the ICNT would appear to possess characteristics 1, 2, 5, and 6 above. First, the organization obviously will deal with specialized and technical subject matter as it will be involved nearly exclusively in deep seabed mining. As such, it will have a clear and unambiguous mission, thus fulfilling characteristic 2. Concerning characteristic 5, it is difficult to determine at this time how large the secretariat would be. Presumably, staff size will be a function of the funds available, and there are no solid figures as yet on which to base such estimates. For characteristic 6, as stated earlier, little public attention has thus far been focused upon the negotiations to create the Authority; "manganese nodules" is hardly a household term.

The projected Authority begins to depart from what Skolnikoff would term an effective organization when characteristic 3 is raised. Membership in the assembly would be universal and the thirty-six-member council reflects the trend of expanding executive bodies. Even more serious is the basis upon which most nations would be chosen, i.e., equitable geographical representation. This standard does not take into account the differing interests and levels of expertise that various nations bring to seabed mining. This issue also has direct relevance to characteristic 4, which states that organizational structure should reflect the interest, power, and knowledge of member governments. Finally, for condition 7, althought the subject is only of moderate economic interest globally, it has, as we have seen, occasioned considerable political interest.

In summary, the Authority in the ICNT would fulfill some of Skolnikoff's criteria for an effective international organization, but by no means would it embody all. The most serious disparity appears to be the fact that the Authority would not be structured to reflect the interests of those nations with the ability to actually mine.

The validity of Skolnikoff's criteria can certainly be questioned. Jacobson has raised a fundamental point that Skolnikoff, for the most part, ignores, namely, that organizations cannot be evaluated without paying attention to the values being expressed through the organization. Jacobson states, "Allowing technology to advance relatively rapidly is an important value and one that presumably will eventually benefit the entire world. Skolnikoff's seeming preference for creating new organizations in which technologically advanced states have a predominant voice would probably give maximum stress to this value. Yet, following such a strategy might risk, at least in the short run, exacerbating the already sharp division between the world's rich and poor."[6] His comments appear particularly relevant to the Authority, because the inspiration and devotion to the establishment of this organization stem not from the desire to further a pioneering industry, as such, but to ensure that the resources gathered from the seabed benefit all mankind (and particularly the developing nations). Given this unique approach and perspective, it is not surprising that the Authority would not resemble a model international organization.

Despite this major caveat to Skolnikoff's criteria, one simply cannot ignore the obvious fact that the Authority envisioned by the ICNT would be seriously handicapped in fulfilling its operational role within the specific functional task of mining. Borgese has quite correctly pointed out, "One of the great challenges in building the International Seabed Authority is to find a new synthesis of economic and political processes."[7] The United States, on the one hand, has emphasized an Authority that would be economically efficient; developing nations, on the other hand, have sought an Authority that would be politically strong. Thus far, no acceptable synthesis of these two goals has appeared. What follows is this author's attempt to describe a synthesis which if seriously considered could provide an acceptable basis or framework for ultimate accommodation.

A Transitional Regime

A truly international solution to the deep seabed controversy can be achieved only if the U.S. administration, Congress, and indeed, the public are willing to concede that times have changed and that the United States can no longer expect to have economic hegemony over the international community indefinitely. Just as the end of political colonialism was inevitable, so too is the end of economic colonialism. If and when this realization arrives, reconciliation should not be difficult to achieve; but without it nothing will be changed.

The basic premise of this chapter is that U.S. policy-makers need not waive basic interests to reach a seabed accord. These policy-makers need only redirect their terms of reference and investigate how *both* U.S. and G-77 interests can be incorporated into a treaty. With this perceptual framework, and some thought and imagination, devising an acceptable formula should not present an insurmountable task.

Similarly, G-77 members will also have to take cognizance of the extent to which the United States and other industrial nations can move immediately toward meeting their demands. The call for a new international order implies a fundamental restructuring of economic and political relationships, and even with the best intentions, political leaders in the West would only be able to slowly move their societies forward. In short, a new international order cannot be generated overnight, and any attempt by G-77 nations to do so will lead to the defeat of their long-range goals.

The existing or current era in which the larger international community finds itself at the moment is neither the post-World War II era, when U.S. economic hegemony was both prevalent and accepted, nor the new international economic order, which has been so loudly proclaimed over the past few years. Whether an era closely resembling the NIEO framework will ever evolve is also the subject of speculation. All that can be stated with certainty at this time is that we are in a transitional period; and whereas it is clear where we have come from, it is decidedly less clear where we are headed.

Thus it would seem to make a good deal of sense to first structure a seabed regime that is also transitional in nature, i.e., a regime that neither reflects the old era nor what G-77 nations would call the new. This goal can best be accomplished by structuring a phased regime in which the Authority would be established to change through the years. In this way, the organization can be specifically set up to adapt with the times.

Given the deadlock over the nature of the Authority and the mining system, it is surprising that very little attention has been devoted to the possibility of this kind of transitional, or interim, international organization. The attractiveness of an evolving, provisional, or phased organization, structured to reflect institutional change over time, stems from its ability to accommodate conflicting objectives. In other words, the formation of an Authority need not be a zero-sum game, creating one winner and one loser. Through a phased Authority, all groups can be accommodated, though it will require deferred gratification for some. The important requirement is that all parties see their essential interests being met—if not now, then sometime in the future.

What is being suggested is that UNCLOS negotiators refrain from separating the Area into various jurisdictional units (as proposed in the banking or parallel systems) which would inevitably produce unequal mining efforts and a refutation of the principle of the common heritage of mankind. Instead of this split in space, a separation in time—through phased regimes—would more appropriately meet the perceived requirements of all parties.

The idea of an evolving Authority has not been completely absent from the negotiations. Charney reports that Committee I Chairman Engo circulated an informal text at the 1975 Geneva session which proposed a phased regime.[8] The first phase generally would have allowed open access to nodules (to existing mining entities) for a period of twenty to forty years. Access in the second phase would then be based solely on the Authority's discretion. This proposal never was incorporated in the SNT.

During the 1976 New York UNCLOS sessions, Kissinger proposed that periodic review conferences of the Authority be held at intervals of perhaps twenty-five years. With this proposal, there would be no promises or assurances that the Authority would in fact change but simply that the opportunity for change would exist.

The ICNT, Article 153, called for a review conference to be held twenty years from the commencement of the seabed regime. Significantly, Article 153(6) stipulates that if states cannot at this review session reach agreement upon a revised regime in five years of negotiations, total control over access and mining conditions will be given to the Authority. This plan definitely goes far beyond what Kissinger was proposing and what the United States now feels is desirable.

Hence, whereas agreement has been reached upon the desirability of a review conference at some time in the future, the terms of reference are still disparate. Far more effort and thought must be devoted to structuring an evolving regime.

INTELSAT

It may be worth looking briefly at a successful international organization which already exists, has broad operational powers in a functional field, and was initially structured to change over time—the International Communications Satellite Organization (INTELSAT). The possible relevance of the INTELSAT model to seabed negotiations was first raised publicly in 1972 by Larry Fabian at the Law of the Sea Institute proceeding.[9] For whatever reasons, however, it has yet to receive careful analysis and consideration.

INTELSAT was formed in 1964 by nineteen nations to establish the first global satellite communications system. It was formed outside of the U.N. system and was intended as strictly a commercial venture. As an organization born in the 1960s rather than the 1970s, INTELSAT has never been charged with the broad mandate (the common heritage of mankind) of the Authority.

When INTELSAT was originally formed, it reflected the hegemony of American technology and expertise in communication satellites. Under a system of representation whereby national voting power was apportioned according to national investment in the space segment, the United States held no less than 61 percent of the votes in the single policy-setting organ, the Interim Communications Satellite Committee (ICSC). Moreover, operational management of the organization was delegated to the U.S. enterprise, the Communications Satellite Corporation (COMSAT), which was responsible for the design, development, construction, establishment, operation, and maintenance of the space segment.

INTELSAT, therefore, was established with something less than a true international flavor.

A key provision in its charter was the Agreement Establishing Interim Arrangements for a Global Communications Satellite System (hereafter referred to as Interim Arrangements). Article IX stated specifically that the accord was in fact a "permanent agreement contemplating revision—but also capable of being continued intact." The Interim Arrangements stated that consideration for specific revision was to be conducted in 1969 and a definitive agreement drawn up by January 1, 1970.

We see, therefore, that INTELSAT was deliberately designed to function in its original years under a provisional regime and structure; i.e., the Interim Arrangements were simply preliminary to the establishment of definitive arrangements, pending an evaluation of operating success.

By nearly all criteria of measuring organizational success, INTELSAT ranked high over its first five years of existence, from 1964 through 1969. The technical feasibility of constructing a global network of communication satellites was proved beyond a doubt, and succeeding generations of satellites with increasing capabilities came on line. The increasing power of communication satellites actually led to the enviable situation of consistently decreasing rate charges to users. These decreasing costs brought new markets and increasing membership; by 1969 some seventy nations were represented in INTELSAT by their various communication entities. Despite this measurable success, when 1969 arrived there was considerable sentiment to alter the structure of INTELSAT as operated under the Interim Arrangements. Hence, as specifically allowed for, negotiations to draw up definitive operating rules were willfully entered into by nations desiring institutional change. After over two years of intensive negotiations, Definitive Arrangements emerged and were initialed on May 21, 1971 (and entered into force on February 12, 1973, with eighty-one nations as signed participants).

Under the Interim Arrangements, INTELSAT had consisted essentially of two structures, COMSAT as the manager and the governing ICSC. Under the Definitive Arrangements, INTELSAT was substantially restructured into a multitier organization with the following organs:

1. An *assembly of parties* considers questions of general policy and has the power to make policy recommendations. Decisions are made on a one-nation, one-vote basis.
2. A *meeting of signatories,* composed of member telecommunication institutions, is charged with providing recommendations on certain operational, financial, and technical aspects of INTELSAT operations. A one-representative, one-vote decision-making rule is also provided for in this body.
3. A *board of governors* forms the executive organ of INTELSAT. This board is the successor to the ICSC and is made up of approximately twenty

members. Voting in this body, composed primarily of industrial nations, is weighted according to national use of the satellite services. Under the ICSC, COMSAT had veto power because of its dominant use of satellite facilities; but under the Definitive Arrangements, a complicated formula was derived whereby no three members, regardless of their weighted votes, could veto an action supported by all other members of the board.

4. A *director general* was to be appointed no later than December 31, 1976, to take over the COMSAT managerial functions, both administrative and technical. Until that time, an appointed secretary-general was to manage administrative, financial, and legal aspects of the operation, with COMSAT retaining technical and operational management.

The internationalization of INTELSAT, provided for by the Definitive Arrangements, has not impeded its functional performance. Costs to users, per unit, continue to decline as satellites become more powerful and additional use is made of satellite circuits.

The purpose of this extended examination of INTELSAT is not to suggest that its organizational structure can or should be transferred intact to that of an Authority. Obviously, the structural differences between INTELSAT and the Authority, as proposed by the ICNT, are considerable. It is difficult to conceive, for example, of any compromise among nations at UNCLOS that would leave the United States with a full 33.6 percent weighted vote in any decision-making organ of the Authority (which is the current voting power of the United States in INTELSAT's board of governors, down from 61 percent under the Interim Arrangements as previously stated).

In my judgment the relevant lesson to be gained from INTELSAT is that an *evolutionary approach* to the formation of international organizations can indeed bring desired change over time, and it might avoid the deadlock that arises from pursuing either a traditional or a revolutionary approach. INTELSAT founders showed considerable foresight in perceiving that an international organization, formed to build a global system around an unproven technology, might prove inadequate over time, either for reasons of success or failure. The firmly established timetable for revision in INTELSAT's charter ensured nations that structural revision was not only possible but anticipated. An essential time period was provided in which to evaluate the performance of the organization and to assess the satisfaction of individual members. In this fashion, revision could be based on existing on-going operations and not on forecasts.

A Phased Authority

One often forgets, in the passion of the argument, that we are dealing with a pioneer enterprise. Not a single commercial mineral from a manganese nodule has yet reached the marketplace. How much it is going to cost to get it there is still open to speculation. Both the technical and economic feasibility of commercially mining manganese nodules remains subject to confirmation.

Given this state of affairs, the most desirable approach might be to reduce the number of roadblocks facing first-generation seabed miners in order to adequately assess the economic viability of this new undertaking. Though this mining could be carried out under the general direction of an Authority, the Authority or its operating organ (the Enterprise) would not directly participate in it. The Authority would play only a service-facilitating role in its initial phase, with operational decisions—outside those agreed to in a treaty—resting primarily with mining enterprises and their sponsoring states. The Authority need not have control over prices and production, since output from the seabed during the first years of commercial mining is likely to be minuscule in comparison to land-based output (with the possible exception of cobalt, which might require some form of specific production-control mechanism). Various mechanisms for compensation could be established for land-based mineral producers as insurance against negative effects. In other words, the initial phase of the Authority—approximately ten to fifteen years—would (1) allow sufficient time to test the feasibility and profitability of the ocean mining industry; (2) allow officials, on the basis of experience, to more intelligently structure a subsequent stage of the Authority's operations; and (3) generate funds to provide a firm financial backing for the Authority.

The developing nations could accept this initial phase only if they are assured—in the treaty itself—that they would have greater control in the subsequent phase (or phases) of the Authority's existence. Definite assurances must be tendered to guarantee that eventually seabed mining will be a truly international undertaking on behalf of the common heritage of mankind. The details of subsequent phases of the Authority need not be spelled out in depth. In fact, for the purposes of flexibility, details should not be included. There must, however, be a clear indication of the desired direction of change.

This is not to say that the principle of the common heritage of mankind should be shelved during the initial phase. Developing nations could be widely represented throughout the Authority, and, most important, there could be a vigorous transfer of technology so that the Authority would gradually garner the skills and know-how to participate directly in mining. The initial phase, therefore, could be viewed by the developing nations as preparation for the much greater responsibilities required in the next one.

Phase II would initiate direct Authority control and participation in mining. First-generation mining operations, initiated under Phase I, would not be affected by Phase II provisions. New mining efforts, however, would be left to the Authority's discretion as to a joint venture, service contract, or any other arrangement. To provide assurances to miners, the Authority would not have the jurisdiction to prevent national or private mining companies from exploiting the deep seabed. It would, however, retain direct and effective control over the nature or pattern of all seabed mining and be allowed to conduct a designated percentage of exploration and exploitation on its own. Surely, the seabed is sufficiently vast and the nodules so diffuse that all interested mining entities, either private, national, multinational, or international, can be accommodated in the foreseeable future.

Beginning in phase II, national membership and voting powers in the key organs of the Authority would be based primarily on achieving regional balance. The Authority would also have limited and carefully defined authorization to implement price and production controls if technical experts determined that developing nations that were land-based mineral producers were being adversely and significantly affected by seabed mining. A revenue-sharing plan, based on any number of taxation methods, could be implemented during this phase, which would not only finance the Authority but also raise development funds for distribution to the poorest nations.

The evolving international organization just described is designed to accommodate the divergent interests expressed at the seabed negotiations. But would developing nations be willing to see these "common heritage" resources exploited during phase I with a minimum of international control? The answer would probably be yes only if clear legal assurances were given that the second phase would begin at a definite future date—which could be accomplished by including a framework for phase II in the treaty. The details could be worked out subsequently.

Another legitimate question is, Would mining enterprises be willing to conduct seabed mining beyond phase I? The answer probably depends on whether first-generation miners make a reasonable return on seabed mining in Phase I. If not, it is quite certain that there will be no great rush to get involved in Phase II. If, on the other hand, miners make a reasonable profit, there is no reason to believe they would not, given the proper assurances, be willing to participate.

Conclusion

Earlier in this chapter it was posited that the deep seabed issue is, in a sense, analogous to that of the Panama Canal. Each represents an issue rich in symbolism that strongly influences the American reaction or response. It is interesting to note that U.S. Panama Canal negotiators produced a legal convention in which time played a key role. Recognizing that it was politically unfeasible to turn the Panama Canal over to the Panamanians immediately, U.S. negotiators instead prepared a treaty that ceded sovereignty of the canal to Panama by the year 2000 while retaining clear rights of international passage. This formula preserved U.S. interests in free passage but demonstrated cognizance of the changing nature of internaional relations and the limits of U.S. power abroad. It is, as stated previously, a movement toward placing the United States within the international community and not on a precarious perch atop it.

Why U.S. seabed negotiators have not chosen a similar path is puzzling.

Surely, ceding sovereignty over an area of strategic importance represents a larger national concession than ceding of operational authority to an international institution mining in an area recognized by all as beyond any national jurisdiction. In both cases, the rewards from resolving these issues far outweigh the sacrifices.

It is very possible, though one cannot state this with certainty, that the G-77 solidarity demonstrated during the negotiations will not carry over into the actual operation of an Authority. Once the enterprise is launched, national perceptions of interest and indeed of the issue itself will probably change. To many nations the functional task of mining has remained in the background to the larger issue of world politics. Once the venture is initiated, however, full attention will likely turn to the task at hand.

By implementing a phased regime with relatively open access to mining sites during the initial phase, the pioneer mining organizations, which have since the 1960s undertaken significant technical development and prospecting, would be able to fully develop and utilize their technology. The penalty they would have paid for having to delay actual commercialization, therefore, would not be too large a burden. Including the framework of subsequent phases in an UNCLOS treaty now would allow mining organizations to make long, lead-time decisions on the basis of explicit and clearly formulated conditions. Regulatory and operating certainty should be the key features of a future seabed regime.

In summary, one can find means to incorporate the interests of all involved. There are no parties under the system just described that would obtain all they want; but neither would there be losers as such. A phased Authority could serve as a model international organization, in the sense that it could illustrate one way to handle the transition from one economic era to another.

Notes

1. Ernst B. Haas. and John G. Ruggie, "International Technology and International Action: What, Who, How and Why?" in *Priority Research Needs on Technology-Related Transnational and Global Policy Problems,* Center for International Studies, MIT, 1973, p. 105.
2. Pardo states that UNCLOS has labored long to accommodate and harmonize the interests of states, and as such has failed to come to grips with the essential management issues. He says that negotiators "prefer to reach advantageous short-term political accommodations which will enable coastal States to digest in peace 35% of the globe (but which will be a disaster in the long term) than to make the political, intellectual, and imaginative effort required to construct a legal framework for man's activities in ocean space that would give us a credible tool for resolving the emerging grave complexes of problems in the oceans." (A. Pardo, "Preliminary Analysis of the 1975 Single Negotiating Text on the Law of the Sea," Woodrow Wilson International Center for Scholars, Washington, D.C., Sept. 1975, p. 101.)

3. William Safire, "Very Deep Thoughts," *New York Times,* July 4, 1977, p. A-17.
4. Christopher W. Pinto, "Problems of Developing States and Their Effects on Decisions on Law of the Sea," in *Law of the Sea: Needs and Interests of Developing Countries,* 7th Annual Conference of the Law of the Sea Institute, June 26–29, 1972, p. 13.
5. Eugene B. Skolkinoff, *The International Imperatives of Technology,* Research Series #16 (Berkeley: Institute of International Studies at the University of California, 1972), pp. 159–60.
6. Harold K. Jacobson, "Technological Development, Organizational Capabilities, and Values," *International Organization,* 25, 4 (Autumn 1971), 782–83.
7. Elisabeth Mann Borgese, "The New International Economic Order and the Law of the Sea," *The San Diego Law Review,* 14, 584 (1977), 589.
8. Jonathan I. Charney, "The International Regime for the Deep Seabed: Past Conflicts and Proposals for Progress," *Harvard International Law Journal,* 17, 1 (Winter 1976), 40–41.
9. Larry Fabian, in *Law of the Sea: Needs and Interests of Developing Countries,* 7th Annual Conference of the Law of the Sea Institute, June 26–29, 1972, p. 92.

Index

Index